SAN JUAN ISLAND LIBRARY DISTRICT
3 3186 00092 4371
P9-DMK-447
DISCARD
5.00

THOREAU'S COUNTRY

THOREAU'S COUNTRY

JOURNEY THROUGH A TRANSFORMED LANDSCAPE

David R. Foster

Harvard University Press · Cambridge, Massachusetts · London, England · 1999

Printed in the United States of America

Illustrations by Abigail Rorer

Library of Congress Cataloging-in-Publication Data

Foster, David R., 1954–
Thoreau's country : journey through a transformed landscape /
David R. Foster.
p. cm.
Includes bibliographical references and index.
ISBN 0-674-88645-3 (alk. paper)
1. Thoreau, Henry David, 1817–1862—Homes and haunts—New England.
2. Foster, David R., 1954– —Journeys—New England. 3. Authors,
American—19th century—Biography. 4. New England—Description and
travel. 5. Landscape changes—New England. 6. Natural history—New
England. I. Thoreau, Henry David, 1817–1862. II. Title.
PS3053.F66 1999
818′.309—dc21
[b] 98-39531

To Marianne

Contents

Preface

In the summer of 1977 I traveled from the place where I had grown up, in the rolling farmland of southern New England, to the woods of northern Vermont with a few tools and the ambition to build a cabin with my own hands and the natural resources that surrounded me. Knowing that I would be living for many weeks without visitors in a temporary shack located a few miles from the nearest road, I carried along a handful of books for company and inspiration. Included in this stack of reading were the journals of Henry David Thoreau. Over the next few months I would occasionally thumb through his writings while sitting on my shaded doorstep or on a rock soaking up sun along the pond shore, but I inevitably put them down with a sense of disillusionment, feeling that the passages were quite far removed from my daily experience in the woods. Thoreau's writings, even those from his own days of building a cabin at Walden Pond, did not mesh with the quiet and continuous forest that surrounded me or with my deep sense of solitude and loneliness.

One Sunday as I was catching a brief rest after a morning spent chopping wood, the faint tolling of a bell reached me, having wound its way up some four miles through the trees and hills from the village located in the valley below. This faint sound of civilization brought a recollection of Thoreau's journals and the regularity with which their passages described the sounds of church bells, locomotives shrieking across the far end of his pond, wagon wheels on the Lincoln road, and townspeople wandering through the woods of Walden. I was struck by the sudden recognition that the sights and sounds and nature that Thoreau encountered on his daily walks through the nineteenth-century countryside of Massachusetts were not those of the deep forest that I was currently living in, but rather more like those I had experienced as a child in the agricultural hills of Connecticut. The area around my family's home was a landscape of corn fields, cow-filled pastures, and small woodlands, each separated by stone walls, and it was filled with the seasonally changing activities of farmers and their families. The smells of fresh-cut hay and recently spread manure, the sights of fencerow trees and neighbors toiling endlessly at their work, the activities of woodchucks, barn swallows, crows, and foxes, and the forays by children into the fields and swamps and orchards and streams were dominant features in this landscape. These memories resonated deeply with Thoreau's daily observations.

My reflections on Thoreau's writings and on these two landscapes in my life raised many questions about the history of New England and the observations of one of America's great nature writers. As my timber felling progressed through that autumn, and as the cut and peeled spruce logs accumulated on the hillside in northern Vermont where I was building my cabin, I had ample opportunity to pause and puzzle over the landscape of Thoreau, to wonder why the writings of a man known for his strong ethic of preservation and wilderness values seemed so foreign to me in the deep forest and yet so alive in a landscape of pastures and cows. In particular, I thought about the nature of Thoreau's woods, which he described as being very open and tame, filled with woodchoppers at their work, children gathering chestnuts or sassafras shoots, and early-rising farmers traveling to market. At the same time, I began to explore my own woods in Vermont and uncovered some surprises.

A wall of stones covered with ferns and moss climbed the forested hillside only a hundred yards from the site that I had chosen for my cabin. Adjacent was a vague, overgrown laneway that cut obliquely across the slope, filled in with a thick cover of striped maple and young spruce. On the other side of the pond, heaps of stones piled four or five feet across and a few feet high were scattered up the hillside, where the dense balsam fir and red spruce nearly obscured them with a thick layer of needles. Next to the stony bubbling spring where I collected water every morning was a deep depression filled with stone and brick and charcoal that marked an old cellar hole. The woods most certainly had a history that I had been quite unaware of, a history of people working and transforming the land, and creating a landscape very different from the one I now occupied. Despite the steep and rocky nature of the land, and the dark extent of the current forest that nearly obscured its history, this whole area had once been inhabited, cleared of trees, and farmed. The more I searched on my daily walks, the more evidence I discovered of this past, in stone walls, wells, and cellar holes scattered through the woods, in bits of rusty barbed wire grown over deeply in a tree, and in the very shape of trees and the patterns they formed across the landscape. The low, spreading branches of a massive sugar maple, now engulfed by a continuous forest of straight and tall younger trees, indicated that in its past it had grown singly along a laneway and had offered shade to cows in an adjoining pasture. On close inspection, the dense hillside stand of red spruce and balsam fir reflected in the pond was strikingly rectangular in outline and evidently represented the first new generation of trees in a field that had been abandoned by some disillusioned farmer nearly a hundred years earlier.

Gradually I perceived the connection between the fields of Connecticut, the passages in Thoreau's journals, and the woods filled with ancient artifacts that now covered northern Vermont. Across New England and much of the eastern United States, the land had once been tamed and farmed through a relentless clearing of forests, piling of rocks, and toiling in fields and barns. Born in 1817, Thoreau had lived through the peak of this agrarian splendor, during a time when the population was spread quite evenly across the landscape in small farming villages and most people made a living by working products from the land. He built his cabin at Walden Pond in one of the few forest areas remaining in Concord, actually a woodlot that served for a century and a half as a source of fuel and timber and food for an active country population, and his journals resound with the sights and sounds and natural history of an open, vibrant, and quite domesticated landscape. Toward the end of Thoreau's life a second transformation of the countryside began, this time back to a wilder state, as New Englanders moved into the new farmlands of the Midwest or relocated into the emerging cities of the Industrial Revolution. As people left the land, pines and birches and blackberries crowded into the neglected pastures and forests sprang up in the old corn fields. Many of the homes, barns, schoolhouses, and wooden fences collapsed, leaving only their more durable remnants in woods that are now crisscrossed with trails descended from an abandoned network of rural roads. The fields that I grew up in, like those of the few other farm landscapes of modern New England, were the meager remains of a once vast countryside of pastures and hay fields and barns that extended from Long Island Sound through Connecticut, Rhode Island, and Massachusetts to northern Vermont, New Hampshire, and southern Maine. Since Thoreau's day, most of the land has reverted to forest, like the one where I was retreating to build my cabin.

The changes in the natural world that I discovered through this reading of Thoreau and my reflection on the New England landscape led me to the realization that nature can only be understood through an awareness of its history. The red spruce that now form the snug walls of my log cabin grew on that Vermont hillside as a consequence of nineteenth-century farming activity and the subsequent decision by a Yankee farmer to quit the area nearly a century ago. The trees that have replaced the spruce as I cut them down are of completely different species, mostly sugar maple, beech, and yellow birch, because this time around nature is filling small openings in the woods, not overtaking large abandoned fields. Meanwhile, the deer that were once so common on my first childhood trips to northern Vermont have mostly disappeared, replaced by a rapidly expanding moose population that thrives in the old, deep forests and wetlands that increasingly characterize New England.

The threads of this and other stories of change extend across New England, the eastern United States, and much of the globe. For the remarkably dynamic history of New England is certainly not a unique story. From the boreal forests of Alaska, where natural forest fires create an endlessly changing mosaic on the land, to the tropical landscape of the Yucatan Peninsula, where the enigmatic decline of a thriving Mayan population a thousand years ago allowed forests to engulf the temple-remnants of a great civilization, the natural world is characterized by change. Understanding this change and interpreting the history of nature are crucial if we are to appreciate, conserve, and manage these natural ecosystems.

After I left my cabin in the woods, as I traveled and studied, I continued to read Thoreau's journals and to walk his landscape on each trip back to New England. Eventually, I was able to return for good to study the land, its ecology, and its history and to use the landscape as a textbook for teaching ecological principles and the complex story of the interactions of humans and nature. In this work I have found Thoreau's daily writings in his journal to be an invaluable source of insight and inspiration. In contrast to *Walden,* which Thoreau edited so rigorously that it is suprisingly free of natural history observations, the journals are full of anecdotes and extended descriptions of the land, its fields and forests, and its people. They are also filled with Thoreau's musings on the importance of nature's history, notes from his own research on changes in his landscape, and reflections on how these changes affected everything he saw.

My aim in writing this book has been to illustrate how dramatically nature can change in a very short period of time and how essential it is to recognize this potential for change in order to appreciate and understand natural landscapes. As he set out to interpret his own landscape, Thoreau used many clues, such as the history of a tree captured in its growth rings, the age of a stone wall determined by the size of lichens growing on it, or the evidence contained in old maps, deeds, and histories, and thus a second objective of this book was to use Thoreau's writings to highlight some of the many approaches one can take to uncover nature's history. Finally, I have used Thoreau's descriptions of the sights and character of New England at its peak of agricultural activity in order to interpret the landscape history that has shaped our modern countryside.

Although Thoreau's descriptions center on Concord, Massachusetts, I did not set out to tell the story of a single township or landscape, nor to assess the scientific abilities or literary development of a single individual. Rather, I am seeking to outline a much broader story about the dynamics of landscapes and the importance of recognizing nature's own history when interpreting and conserving it. Thoreau's writings shed light on the importance and application of this historical approach, and thus a substantial part of each chapter consists of Thoreau's own words—excerpts from the journals that convey his appreciation for landscape change and provide insights into the processes that shaped his landscape and have led to the development of ours. My own observations are intended as a companion to Thoreau's, to provide a context for his descriptions and in some cases to explain them more fully. I also discuss the ways in which Thoreau's observations may be applied to the understanding and conservation of modern landscapes.

This volume owes its development and inspiration to many people. I thank Anne and Pete Foster for a rural upbringing in a Connecticut landscape that retained a surprising resemblance to Thoreau's in appearance and activity and for encouraging me to experience the northern, wilder part of New England as well. My understanding of New England land-use activity began on the Bartholomew, Cella, and Anderson farms, where I worked and learned, among other things, to pick stones from plowed fields and then to spread the manure; I also benefited from the friendship of Bill Tresselt, who introduced me to squirrel hunts, bass fishing, muskrat traps, and more, and that of Jim Gaffey and then Fred Keogh, who joined me for many exploits. From Bill Niering, Dick Goodwin, and Betty Thomson I learned to interpret quite a bit of landscape history, and with Herb Wright and Bud Heinselman I began to put this ability into practice. Much of my understanding of the ecology and history of the New England countryside developed at the Harvard Forest, through discussions with Ernie Gould, Hugh Raup, Glenn Motzkin, and John O'Keefe, and from many walks with classes and colleagues. Gordon Whitney and George Peterken inspired me to put this ecological understanding in a rigorously historical context; Barbara Flye has made the development of this book an ongoing and delightful collaboration; and Dorothy Smith has helped to put all of my work in final form.

The research behind this project has been underwritten by the A. W. Mellon Foundation and the National Science Foundation, which have been extremely supportive of studies of landscape change. The completion of this manuscript has benefited greatly from the comments, suggestions, and encouragement of Marianne Jorgensen, Pete and Anne Foster, Jim Gaffey, Glenn Motzkin, John O'Keefe, Barry Tomlinson, Larry Buell, Kay Gross, Harry Foster, Birgit Jorgensen, John Burk, Kate Scott, and Ray Angelo. Abbie Rorer's shared interest in and familiarity with the history of the New England countryside have made working with her on the illustrations a great pleasure. At Harvard University Press the support and advice of Ann Downer-Hazell, Mary Ellen Geer, and Michael Fisher have greatly improved this book. Most important, my first question concerning the name and biology of a tree (it was a very shade-tolerant beech) was answered by Marianne Jorgensen in 1970. She has inspired me to follow this question with countless others and has joined with me and Christian and Ava in many wonderful pursuits. For this I will be forever grateful.

THOREAU'S COUNTRY

Prologue:
One Man's Journal

"What are you doing now?" he asked. "Do you keep a journal?" So I make my first entry today.

OCTOBER 22, 1837

In the Morgan Library in New York City there are thirty-nine simple journeyman's books, of varying size and shape, neatly bound and strapped in a pine case handcrafted by the author himself. Written in a flowing though nearly illegible hand, they contain the thoughts and observations of a man who spent all but a few of his years roaming the nineteenth-century countryside of Concord, Massachusetts. Although he traveled, he was a man who recognized that a universe existed within himself and the borders of his own landscape—a universe of field and forest; of meadow and stream; of birds and mammals and domesticated livestock; of people living and dead, whose activities and toil he admired, and whose influence on his countryside intrigued him.

The writer was Henry David Thoreau. By the time that tuberculosis stilled his thoughts and pen in 1862, the journals had grown to more than two million words, a remarkable legacy of writing and observation. By this time, as well, the New England landscape had been transformed from the wilderness of Indians, old-growth forest, and bountiful wildlife that greeted the early colonists to an agricultural countryside of farmers, meadows, woodlots, and meadowlarks. Since Thoreau's death this landscape has been transformed yet again: the farmers have left their lands, trees and woodland plants have taken over the fields, and a forest filled with wildlife has appeared once more. Thoreau's journals therefore provide more than the literary and philosophical gems for which they are commonly admired and mined. These daily observations are a priceless source of information about the changing relationship between people and the natural environment that has shaped the land and continues to determine where it is heading.

As the quotation at the beginning of this chapter reveals, Thoreau began his habit of regular journal entries in 1837, when he was twenty, after Ralph Waldo Emerson suggested that he use a journal to develop his skills of writing and observation. Writing nearly daily and composing frequent lengthy passages, he maintained this routine until his death. The journals, along with his separate volumes of quotes and lessons from books that he read, represent literally his life's work, for Thoreau considered himself to be fully employed with his responsibilities of experiencing nature, recording his thoughts and details of natural history, and then combining the resulting observations and reflections into the daily journal entries. Only a small subset of these notations were subsequently extracted and carefully edited by Thoreau to form the two volumes that appeared during his lifetime, *Walden* and *A Week on the Concord and Merrimack Rivers.*

Thoreau's insights into the New England landscape and natural history are the product of his single-minded devotion to the daily observation and interpretation of the countryside that surrounded him and his commitment to understanding the history of changing forces that shaped everything he saw. His journal is a rich source of information precisely because he lived in and described one relatively small part of New England in the context of its changing history. Rather than choosing to explore the world and document nature and human activity extensively, Thoreau sought to reflect intensively on the daily, seasonal, and annual happenings of the New England countryside. The result is an unparalleled wealth of observations and reflections that were based on repeated visits to the same sites, under changing conditions, separated by time and the opportunity for careful thought. Thus he was able to document the commonplace as well as interpret the complex.

Thoreau pursued this approach of intimate observation and careful reflection because he was convinced that all he needed to learn of importance was inside himself and that much of this could be brought out and illuminated through observations of his surroundings. This belief persuaded him to stay home to explore himself and nature. Through this process he found remarkable local analogues to a wide range of human experience and emotion. In order to witness the emotion and struggle of a war, he turned his gaze from Homeric epics to the red and black ants that engaged in anxious, single-minded battle in his cabin dooryard at Walden. To enjoy wild landscapes, he secluded himself in Gowing's swamp, where he became convinced that wildness is more an attribute of man than of nature. To observe human virtue or vice, or to record examples of the power and beauty of nature, he searched the fields, hills, and farmyards of Concord, where he found each in abundance. He seldom felt the urge to travel—after all, he stated, when one travels to a new place, one arrives only to discover oneself.

The conviction that true knowledge and experience come from exploring oneself in the natural world that surrounds us led Thoreau to a comprehensive documentation of the fully humanized landscape of New England. By devoting himself to this activity, he became an observer of people, their impact on the natural world, and the plants, animals, and landscapes that they shaped. Thus, while he often walked across lots to avoid a land owner, he just as frequently found himself observing and admiring the farmers and cattle that he encountered in the pastures. As he traveled the woods to revel in the company of the few remaining grand trees, he found himself absorbed in the tales of woodland history recounted by Alek Therien, a resident woodchopper. As he floated on streams looking at musquash (muskrat), he often ended up following the trapping activities of John Goodwin, the one-eyed fisherman, or sampling the fumes of blackstrap liquor in a jug abandoned by the men who mowed the meadows. Paradoxically, one of America's great nature writers also had a keen interest in people, and he probed deeply into the complex interactions between humans and their environment.

Few days in Thoreau's life passed without some note-taking or a journal entry, and many entries consisted of lengthy passages describing the activity and stories of his countryside acquaintances, or recounting an excursion by foot or boat. Others combined thoughts, observations, and recent readings into a philosophical or historical discussion on the state of man and society. Thoreau's journal entries convey a sense of immediacy that suggests spontaneous invention; however, it is clear that they were crafted with care and great deliberation. Individual entries were based on his jottings in the field and were rewritten from these preliminary drafts. Many events and thoughts appear in different entries on successive days. Writings from one day may be cross-referenced, corrected, or expanded in later entries. Library readings and scientific discussions with experts in botany, zoology, and geology in Cambridge and Boston are quoted and referenced, sometimes extensively. And many passages from *Walden* and *A Week on the*

Concord and Merrimack Rivers appear in earlier form throughout the journals. For Thoreau, writing was a joy as well as an exercise in developing the skills of observation and communication, and it filled his daily life and lifetime.

What need to travel? There are no sierras equal to the clouds in the sunset sky.

JANUARY 11, 1852

It is in vain to dream of a wildness distant from ourselves. There is none such. It is the bog in our brain and bowels, the primitive vigor of Nature in us, that inspires that dream. I shall never find in the wilds of Labrador any greater wildness than in some recess in Concord, *i.e.* than I import into it.

AUGUST 30, 1856

I have given myself up to nature; I have lived so many springs and summers and autumns and winters as if I had nothing else to do but *live* them, and imbibe whatever nutriment they had for me; I have spent a couple of years, for instance, with the flowers chiefly, having none other so binding engagement as to observe when they opened; I could have afforded to spend a whole fall observing the changing tints of the foliage.

SEPTEMBER 19, 1854

If these fields and streams and woods, the phenomena of nature here, and the simple occupations of the inhabitants should cease to interest and inspire me, no culture or wealth would atone for the loss.

MARCH 11, 1856

The question is not where did the traveller go? what places did he see?—it would be difficult to choose between places—but who was the traveller? how did he travel? how genuine an experience did he get?

JANUARY 11, 1852

A man must generally get away some hundreds or thousands of miles from home before he can be said to begin his travels. Why not begin his travels at home? Would he have to go far or look very closely to discover novelties? The traveller who, in this sense, pursues his travels at home, has the advantage at any rate of a long residence in the country to make his observations correct and profitable. Now the American goes to England, while the Englishman comes to America, in order to describe the country. No doubt there [are] some advantages in this kind of mutual criticism. But might there not be invented a better way of coming at the truth than this scratch-my-back-and-I-'ll-scratch-yours method? Would not the American, for instance, who had himself, perchance, travelled in England and elsewhere make the most profitable and accurate traveller in his own country? How often it happens that the traveller's principal distinction is that he is one who knows less about a country than a native! Now if he should begin with all the knowledge of a native, and add thereto the knowledge of a traveller, both natives and foreigners would be obliged to read his book; and the world would be absolutely benefited.

AUGUST 6, 1851

Flag Hill is about eight miles *by the road* from Concord. We went much further, going and returning both; but by how much nobler road! Suppose you were to ride to Boxboro, what then? You pass a few teams with their dust, drive through many farmers' barn-yards, between two walls, see where Squire Tuttle lives and barrels his apples, bait your horse at White's Tavern, and so return, with your hands smelling of greasy leather and horsehair and the squeak of a chaise body in your ears, with no new flower nor agreeable experience.

JUNE 19, 1852

Many a man, when I tell him that I have been on to a mountain, asks if I took a glass with me. No doubt, I could have seen further with a glass, and particular objects more distinctly,—could have counted more meeting-houses; but this has nothing to do with the peculiar beauty and grandeur of the view which an elevated position affords. It was not to see a few particular objects, as if they were near at hand, as I had been accustomed to see them, that I ascended the mountain, but to see an infinite variety far and near in their relation to each other, thus reduced to a single picture.

OCTOBER 20, 1852

If a man walks in the woods for love of them and [to] see his fellows with impartial eye afar, for half his days, he is esteemed a loafer; but if he spends his whole days as a speculator, shearing off those woods, he is esteemed industrious and enterprising—making earth bald before its time.

JUNE 17, 1853

Three Landscapes in New England History

New England and much of the eastern United States have undergone a complete and quite astonishing ecological transformation in the last 350 years. From the beginning of colonial New England, with the establishment of settlements at Plymouth, Boston, Narragansett Bay, and New Haven, settlers spread rapidly across the landscape and set about the business of converting forested wilderness into productive agricultural land. By the late 1700s thriving farms began to dominate the New England landscape, and by the early 1800s more than 60 percent of the land was in open fields, interspersed with small woodlands and crisscrossed by a dense network of roads. The human population was dispersed quite evenly across this landscape, in small agricultural and commercial townships of five hundred to a few thousand individuals. Most people lived off the land and its products, in one way or another. The nineteenth century was the heyday of what has become our romantic image of the rural New England countryside: agricultural villages, small shops and mills, open hillsides, cows and sheep grazing in fields, bell-laden horses drawing sleighs along country laneways, and ranks of farmers mowing hay and grain with long curved scythes.

However, beginning in the 1830s with the expansion of the large linen mills in Lowell, Massachusetts, and the laying of the first railroad lines across New England, an industrial and social revolution was initiated that changed the people and their relationship to the land, and led to one of the most remarkable transformations in world history. In increasing numbers, people abandoned their farming and rural life styles and moved either to the emerging New England cities, to the productive farmlands and timberlands of the Midwest, or to the alluring gold fields in the West, joined in all three destinations by growing numbers of foreign immigrants. As New Englanders deserted their farms for the expanding mill towns, urban centers, and distant destinations, the plowed fields became rough pastures, the abandoned pastures and meadows grew into shrublands and new forests, and collapsed cellar holes, stone walls, and abandoned laneways became frequent sights in an increasingly natural and forested landscape. The little agriculture that did remain was concentrated in the most accessible and productive land: in the fertile soils of the broad valley bottoms or in gentle, rolling terrain. Most of New England reverted back to forest, and the land became wilder.

Today, the landscape of New England ranges from 60 to 95 percent forest. This relatively abrupt transformation of the land from open fields to dense woodland affects people's livelihoods, the natural history of forest, lake, and stream ecosystems, and the changing aesthetic character of the countryside. The regrowing forest has provided the raw material for a series of minor revolutions in natural resource use and appreciation. At the turn of the century a tremendous logging industry developed, based on harvesting of the white pine trees that had grown up on the old abandoned farmland. The wood of white pine was particularly well suited for constructing the containers (pails, baskets, boxes, barrels) that were a part of everyday life in the era before cardboard, plastic, and container ships. A second timber era continues today, centered on the oak, ash, maple, birch, beech, and other hardwood species that grew up on the cut-over pinelands and other second-growth forests. In general, however, the New England countryside outside of northern Maine—and this is true of most of the United States—continues to grow more timber than is cut, and therefore the forests are getting continually older and individual trees are getting larger. Appreciation for this natural (albeit new) forest landscape and the recreational opportunities that it affords has fueled the development of important new ways of looking at and using the land and deriving income from it. Old-growth forests are sought out and cherished when discovered; intensive logging activity as well as trapping and hunting of wildlife are viewed negatively by much of the population; forest and natural area managers at the local, state, and national level are receiving increased demand for "wilderness" experiences and backwoods recreation; and the travel industry and state tourist boards are responding by developing new forms of eco-tourism, ranging from fall foliage tours to cross-country skiing to the bed and breakfast phenomenon.

As New England forests have expanded in area, as the trees have increased in size, and as social attitudes toward nature have changed, the animal life of New England has also been transformed. Within the lifetime of most residents there has been a notable decline in species that thrive in open grasslands, shrublands, and pastures. Bobolinks, meadowlarks, grasshopper sparrows, and woodcock have decreased dramatically in abundance and have become a major focus of new environmental concern. At the same time, residents are increasingly reminded by newspaper accounts and personal experience that the wildlife of forests and deep woods is on the increase, and on the move. Beaver, turkey, bear, fisher, pileated woodpecker, owl, and bobcat have all proliferated, as have populations of deer, coyote, moose, and even eagles. Consequently, while New Englanders enjoy their scenic forests, they wonder how to deal with vehicular encounters with moose and backyard visits by bears, and they lament the loss of farms and farmland along with their associated scenery and wildlife. The rolling fields dotted with cows and the open vistas of the New England countryside have been replaced by woods (or by suburban development), with only the stone walls winding among the trees to serve as reminders of the agricultural and rural past.

Thus, New England is a cultural landscape, shaped by the interaction of human history and the natural environment. Nearly every acre of the countryside has been directly affected by past land use; in fact, most of the modern forest is located on former agricultural land that was grazed by cattle, tilled for crops, or cut regularly to provide a farmer with wood for fuel. The New England landscape is continuing to change as it recovers slowly from these earlier impacts and becomes afflicted with new ones. In order to understand the condition of the land today, to anticipate future change, and to manage nature and preserve it wisely, we need to understand its past. We need to understand the land's potential, its historical appearance, and the specific ways in which people treated particular types of land or individual species of plants or animals. With an appreciation of this history, we can better understand how the modern landscape has developed: why, for example, a crumbling stone wall often separates two distinctly different types of forest; why the morning call of the bobwhite is becoming a distant memory, replaced by the gobble of a clutch of turkeys or the sharp cry of a raven; why the lilacs and red cedars and columbine of open yards and fields are found in dense woods; why our bird feeders are being destroyed by marauding bears; why our suburbs are increasingly serving as host to wandering moose.

As we seek to define and describe the ecological transformation of New

England, a reading of Thoreau's journals provides an insightful perspective originating from a pivotal period in our history. A century before Thoreau's birth in 1817, the native Indian population of New England was decimated by disease and battle and had moved its activities to the new frontier colonies in western Massachusetts. Most of interior New England remained unsettled by European colonists, and the upland regions were as wild as they had been for millennia. In his quest to learn more about the environment and natural history of this early New England wilderness, Thoreau used historical information, his own studies of forest change, and his travels to primitive woodlands undisturbed by humans in order to form his reconstructions of what the early colonial landscape must have been like.

By 1845, when Thoreau set up temporary residence in the cabin of his own making on Emerson's land at Walden Pond, the New England landscape was near its peak of deforestation and was being farmed and used as intensively as it ever has been. In Thoreau's parlance, nature had been tamed; in ours, New England had been transformed into a cultural landscape. As far as one could see the land was directly shaped in most characteristics by human hands or by domestic animals, and it was quite distinct from anything that had preceded it or would succeed it. This was the landscape that Thoreau studied, documented, and reflected on during his daily walks.

Today, more than a hundred years after Thoreau's death in 1862, much of the New England landscape has been reforested naturally; the greatly expanded human population is now concentrated in cities and suburbs rather than rural towns; and most of the energy, food, and materials of manufacture for the eastern United States are derived from outside the region. Thoreau anticipated and documented the forces behind these cultural and ecological changes: industrial development, new socioeconomic pressures, and the opening of the agricultural frontier of the Midwest. He also recorded in great detail many of the processes that gave rise to the forests that we have today. Thus, in his thoughts, his observations, and his very life span, Thoreau captures the essence of transformed New England. We can use the insights recorded in his journal to look back at an earlier landscape largely untouched by European activities; we can read his notes for detailed descriptions of contemporary land-use practices, vistas, and a New England environment that we can never know; and we can apply this information in interpreting the development and current function of our modern landscape.

In Thoreau's writing we can see both the sense of loss he felt in recognizing the extent of environmental transformation that had occurred in the history of New England and the exhilaration of living in a cultural landscape in which people and the environment together generated diverse patterns in nature that changed with every walk and each succeeding day. The wilderness of the Indian was gone, but it had been exchanged for a land in which grain fields, woodlands, and church spires gave distinct character to each set of distant hills; where the air was filled with the pollen of grass, ragweed, sorrel, and pine; where the late afternoons rang with the distant echoes of wheat flailing and wood chopping, with the booming of bitterns. These sights and sounds inspired the passion that shines through all of Thoreau's writing for both the wilderness of the distant recesses of New England and the humanized landscape in which he lived. It is these twin passions that prompted him to advocate the preservation of remaining woodland areas and to decry the rapaciousness of humans, while simultaneously offering advice on how to improve logging practices and lauding the farmer as a noble and heroic model of mankind. Through his journals we receive a wide-ranging appreciation and respect for nature and for nature altered by people, despite the fact that the latter is somewhat at odds with the widespread but misconstrued image of Thoreau as a man of wilderness solitude.

Given that Thoreau was an inspirational nature writer whose words have helped to motivate a modern wilderness and preservationist movement in America, it is often presumed that he lived in a largely natural world. When we look back to his earlier and less mechanized age, we may suppose that mid-nineteenth-century New England was a simpler and perhaps more environmentally intact land than our own, a place where vestiges of the original North American wilderness were interspersed with the bucolic fields of self-sufficient country folk. However, this nostalgic vision is quite incorrect. Walden was just one of many New England woodlots that supplied fuel to households, each of which consumed 10 to 30 cords of wood per winter to heat a drafty home with an inefficient fireplace or stove. Concord was a commercially thriving agricultural community and regional crossroads in which cows and chickens greatly outnumbered native wildlife and meadows overwhelmed forests. Massachusetts was leading a national and global revolution in industrialization and social change. The New England landscape was decried by some as devastated, while others warned it was on the brink of a critical wood shortage.

Thus, in the short period since European settlement, the New England landscape had been transformed. As Thoreau walked through the forest lands of Concord, he noted not their wild condition, but the fact that "every larger tree which I knew and admired is being gradually culled out and carried to mill." When he hiked to the top of a hill for a view, the picture he described was beautiful in its colors and images, but controlled in every detail by the activities of his fellow townsfolk. The panoramas from the Conantum cliff or Fair Haven Hill, favorite vantage points from which Thoreau oversaw his universe, were a patchwork of fields and meadows and woodlots, mostly rectangular in outline and each bounded by "stone walls with post and riders," "rail fences," or "fences and ditches." Each patch had its distinctive color and specific purpose in this extensively worked landscape. Among the sights that Thoreau saw were cleared land that was previously wooded and now awaited its next use as pasture or field; cultivated plowlands with the stubble of grain crops turned under; stream-side meadows of coarse native grasses mixed with flowers and shrubs that awaited the scythe to be turned into winter feed for livestock; sproutlands that supplied the fuel and wood for use on the farm and in town; pine and oak forests that yielded sawlogs for the neighboring mill; ice on the ponds that was cut, hauled, and packed away in sawdust-lined sheds, to be used during the heat of summer; and cranberries growing on the andromeda swamps that were harvested annually for home use.

Thoreau alternately marveled at and lamented these patterns in the tamed landscape of Concord. When he compared his Massachusetts countryside to what he could reconstruct of the earlier forests and conditions of New England, he felt deprived. It was not just the forest that was altered; the wildlife populations had been devastated by the combined impact of changing habitat and relentless slaughter. The extent to which Concord's nature deviated from wildness was abundantly clear and sobering.

However, despite the tameness of the countryside that Thoreau traversed, the open pastoral landscape of New England was home to plant species, rural personalities, habitats, and aesthetic wonders that captivated him and would seem novel to a twentieth-century viewer. The sproutlands, oak and pine plains, and broad meadows of nineteenth-century Concord are largely unknown in modern New England. The open agricultural vistas that Thoreau described on his walks across Massachusetts have been hidden for many years by towering forests. It is the connection that Thoreau forms between the wilderness of early New England, the pastoral agricultural era of the nineteenth century, and the reforested landscape of today that is invaluable to us. Through his musings and reflections on everyday scenes and activities, Thoreau provides us with new insight into the New England countryside and new ways of recognizing and appreciating the history of natural change.

The Cultural Landscape of New England

Views of the Nineteenth-Century Countryside

As I look northwestward to that summit from a Concord cornfield, how little can I realize all the life that is passing between me and it,—the retired up-country farmhouses, the lonely mills, wooded vales, wild rocky pastures, and new clearings on stark mountain-sides, and rivers murmuring through primitive woods! All these, and how much more, I *overlook.*

SEPTEMBER 27, 1852

In order to begin to understand our own landscape, we need to visualize the changes that it has undergone. In particular, it is critical that we become familiar with the appearance of New England in the nineteenth century, when intensive agriculture reigned. The daily descriptions in Thoreau's journals offer striking images of the cultural landscape in which he lived: the open land, the wide vistas, the wonderful patchwork of colors imparted by the diverse agricultural activities of the day.

I sit on the hillside near the wall corner, in the further Conantum field, as I might in an Indian-summer day in November or October. These are the colors of the earth now: all land that has been some time cleared, except it is subject to the plow, is russet, the color of withered herbage and the ground finely commixed, a lighter straw-color where are rank grasses next water; sprout-lands, the pale leather-color of dry oak leaves; pine woods, green; deciduous woods (bare twigs and stems and withered leaves commingled), a brownish or reddish gray; maple swamps, smoke-color; land just cleared, dark brown and earthy; plowed land, dark brown or blackish; ice and water, slate-color or blue; andromeda swamps, dull red and dark gray; rocks, gray.

DECEMBER 23, 1855

Thoreau's New England was a farming and rural landscape that was actively used by a wide range of people to yield a diversity of products. Thus his writings are filled with descriptions of grassy expanses of different types and uses—lowland meadows that produced rough feed, hilly and rocky pastures covered with roaming cattle, hay and grain fields of endless variety that farmers tended and cultivated carefully for winter fodder. Even the pastures themselves were diverse: some were rocky and wild, others were dotted with young pines, still others were expansive, broken by large shade oaks and hickories with their bark rubbed smooth by the hides of grazing cows. From a distance, many pastures had a ribbed and contoured pattern marking the parallel paths worn by the daily passage of cows as they crossed along the slopes in search of grass. Often hidden within the herd or behind the solitary pasture oak was a bull of uncertain temperament, always a potential liability for the unwary walker.

The hay fields were equally varied and colorful, with the deep green of clover, the bright yellow of meadow sedges intermixed with flowers, the pale blue-green of the grains, and the shining appearance of red-top grass.

The next field on the west slopes gently from both east and west to a meadow in the middle. So, as I look over the wall, it is first dark-green, where white clover has been cut (still showing a myriad low white heads which resound with the hum of bees); next, along the edge of the bottom or meadow, is a strip or belt three or four rods wide of red-top, uncut, perfectly distinct; then the cheerful bright-yellow sedge of the meadow, yellow almost as gamboge; then a corresponding belt of red-top on its upper edge, quite straight and rectilinear like the first; then a glaucous-green field of grain still quite low; and, in the further corner of the field, a much darker square of green than any yet, all brilliant in this wonderful light. You thus have a sort of terrestrial rainbow . . .

JULY 22, 1860

This diversity of grasses and the variations of patterns and colors that they made would be quite striking to someone of our day, when simplified, productive, and large-scale agriculture, which tends so strongly toward monoculture and expansive fields, has largely done away with pastures and open grazing in favor of the continuous barn feeding of livestock. In Thoreau's time, when local farmers produced the diverse range of goods necessary to meet most needs of a town, and when hand tools and the power of oxen and horses enabled farmers to adjust their agricultural activity to fit closely the varied topography and soils of the New England land, the landscape patterns that emerged were complex on a fine scale and quite varied. Thoreau's description of the view from the hills of Conantum above Fairhaven Bay in south Concord captures these patterns beautifully, with the juxtaposition of sproutlands, meadow, cleared land, swamps, plowed land, ice and water, andromeda swamps, and rocks. Equally striking are the place names that he uses—Boulder Field, Yellow Birch Swamp, Lime Quarries and Kiln, Hog Pasture, and Ermine Weasel Woods.

In this landscape of scattered forests set against open fields, Thoreau found that it was not always easy on his cross-country walks to screen himself from people and to avoid houses. The Concord landscape was vibrant and beautiful, but its remarkable openness and activity made solitude somewhat elusive.

> What is a New England landscape this sunny August day? A weather-painted house and barn, with an orchard by its side, in midst of a sandy field surrounded by green woods, with a small blue lake on one side. A sympathy between the color of the weather-painted house and that of the lake and sky.
>
> AUGUST 26, 1856

> I can see from my window three or four cows in a pasture on the side of Fair Haven Hill, a mile and a half distant. There is but one tree in the pasture, and they are all collected and now reposing in its shade, which, as it is early though sultry, is extended a good way along the ground. It makes a pretty landscape . . . It shows the importance of leaving trees for shade in the pastures as well as for beauty. There is a long black streak, and in it the cows are collected.
>
> JUNE 20, 1850

A large swelling pasture hill with hickories left for shade and cattle now occupying them. The bark is rubbed smooth and red with their hides. Pleasant to go over the hills, for there is most air stirring, but you must look out for bulls in the pastures. Saw one here reclining in the shade amid the cows. His short, sanguinary horns betrayed him, and we gave him a wide berth, for they are not to be reasoned with.

JUNE 19, 1852

It requires considerable skill in crossing a country to avoid the houses and too cultivated parts,—somewhat of the engineer's or gunner's skill,—so to pass a house, if you must go near it through high grass,—pass the enemy's lines where houses are thick,—as to make a hill or wood screen you,—to shut every window with an apple tree. For that route which most avoids the houses is not only the one in which you will be least molested, but it is by far the most agreeable. . . . We crawled through the end of a swamp on our bellies, the bushes were so thick, to screen us from a house forty rods off whose windows completely commanded the open ground, leaping some broad ditches, and when we emerged into the grass ground, some apple trees near the house beautifully screened us.

JUNE 19, 1852

These are very agreeable pastures to me; no house in sight, no cultivation. I sit under a large white oak, upon its swelling instep, which makes an admirable seat, and look forth over these pleasant rocky and bushy pastures, where for the most part there are not even cattle to graze them, but patches of huckleberry bushes, and birches, and pitch pines, and barberry bushes, and creeping juniper in great circles, its edges curving upward, and wild roses spotting the green with red, and numerous tufts of indigo-weed, and, above all, great gray boulders lying about far and near, with some barberry bush, perchance, growing half-way up them; and, between all, the short sod of the pasture here and there appears.

JUNE 23, 1852

Looking down into the singular bare hollows from the back of hill near here, the paths made by the cows in the sides of the hills, going round the hollows, made gracefully curving lines in the landscape, ribbing it.

AUGUST 3, 1852

A rambling, rocky, wild, moorish pasture, this of Hunt's, with two or three great white oaks to shade the cattle, which the farmer would not take fifty dollars apiece for, though the ship-builder wanted them.

DECEMBER 22, 1852

The need of fuel causes woods to be left, and the use of cattle and horses requires pastures, and hence men live far apart and the walkers of every town have this wide range over forest and field.

APRIL 21, 1852

Haverhill [a small town in eastern Massachusetts] is remarkably bare of trees and woods. The young ladies cannot tell where are the nearest woods.

APRIL 21, 1853

A long row of elms just set out by Wheeler from his gate to the old Lee place. The planting of so long a row of trees which are so stately and may endure so long deserves to be recorded. In many localities a much shorter row, or even a few scattered trees, set out sixty or a hundred years since, is the most conspicuous as well as interesting relic of the past in sight.

MAY 8, 1853

What shall this great wild tract over which we strolled be called? Many farmers have pastures there, and wood-lots, and orchards. It consists mainly of rocky pastures. It contains what I call the Boulder Field, the Yellow Birch Swamp, the Black Birch Hill, the Laurel Pasture, the Hog-Pasture, the White Pine Grove, the Easterbrooks Place, the Old Lime-Kiln, the Lime Quarries, Spruce Swamp, the Ermine Weasel Woods; also the Oak Meadows, the Cedar Swamp, the Kibbe Place, and the old place northwest of Brooks Clark's . . . It is a paradise for walkers in the fall. There are also boundless huckleberry pastures as well as many blueberry swamps. Shall we call it the Easterbrooks Country? It would make a princely estate in Europe, yet it is owned by farmers, who live by the labor of their hands and do not esteem it much. Plenty of huckleberries and barberries here.

JUNE 10, 1853

Nature made a highway [i.e., the Concord River] from southwest to northeast through this town (not to say county), broad and beautiful, which attracted Indians to dwell upon it and settlers from England at last, ten rods wide and bordered by the most fertile soil in the town, a tract most abounding in vegetable and in animal life; yet, though it passes through the center of the town, I have been upon it the livelong day and have not met a traveller.

AUGUST 30, 1853

For a long distance, as we paddle up the river, we hear the two-stanza'd lay of the pewee [a small bird] on the shore,—pee-wet, pee-wee, etc. Those are the two obvious facts to eye and ear, the river and the pewee. After coming in sight of Sherman's Bridge, we moored our boat by sitting on a maple twig on the east side, to take a leisurely view of the meadow. The eastern shore here is a fair specimen of New England fields and hills, sandy and barren but agreeable to my eye, covered with withered grass on their rounded slopes and crowned with low reddish bushes, shrub oaks. There is a picturesque group of eight oaks near the shore, and through a thin fringe of wood I see some boys driving home an ox-cartload of hay.

APRIL 2, 1852

I think our overflowing river far handsomer and more abounding in soft and beautiful contrasts than a merely broad river would be. A succession of bays it is, a chain of lakes, an endlessly scalloped shore, rounding wood and field. Cultivated field and wood and pasture and house are brought into ever new and unexpected positions and relations to the water. There is just stream enough for a flow of thought; that is all.

APRIL 16, 1852

From N. Barrett's road I look over the Great Meadows. The meadows are the freshest, the greenest green in the landscape, and I do not (at this hour, at any rate) see any bent grass light. The river is a singularly deep living blue, the bluest blue, such as I rarely observe, and its shore is silvered with white maples, which show the under

sides of their leaves, stage upon stage, in leafy towers. Methinks the leaves continue to show their under sides some time after the wind has done blowing. The southern edge of the meadow is also silvered with (I suppose) the red maple. Then there is the darker green of the forest, and the reddish, brownish, and bluish green of grass-lands and pastures and grain-fields, and the light-blue sky.

JUNE 23, 1852

Evening . . .

Moon not up. The dream frog's is such a sound as you can make with a quill on water, a bubbling sound . . . The spearers [spear fishermen] are out, their flame a bright yellow, reflected in the calm water. Without noise it is slowly carried along the shores. It reminds me of the light which Columbus saw on approaching the shores of the New World. There goes a shooting star down towards the horizon, like a rocket, appearing to describe a curve. The water sleeps with stars in its bosom.

MAY 5, 1852

The loudest sound produced by man that I hear now is that of a train of cars passing through the town. The evening air is so favorable to the conveyance of sound that a sudden whistle or scream of the engine just startled me as much as it does near at hand, though I am nearly two miles distant from it. Passed two silent horses grazing in the orchard, and then a skunk prowling on the open hillside, probably probing for insects, etc . . .

The chief sounds now are the bullfrogs and the whip-poor-wills. The *er-er-roonk* of the bullfrog actually sounds now without a pause from one end of this river to the other, and can be heard more than a mile on each side. I hear the beat of a partridge also. Is it not a result of the white man's intrusion and a sign of the wildness of the bird, that it is compelled to employ thus the night as well as the day? Though frogs and crickets and gnats fill the air with sound, these horses, great beasts as they are, I cannot detect by any sound they make, but by their forms against the sky. The Cliff rocks are warm to the hand. It is probably after ten. I just came through a moonlit glade in the woods on the side of the hill, where an aspen *(Populus grandidentata)* trembled and betrayed a rising wind. A cuckoo I just heard, an imperfect note, and a wagon going over a bridge, I know not where. It is soon over, and the horse's hoofs and the wheels are no longer heard. That small segment of the arc which the traveller described is remarkably distinguished.

JUNE 18, 1853

Daily Life

From many a barn these days I hear the sound of the flail. For how many generations this sound will continue to be heard here! At least until they discover a new way of separating the chaff from the wheat.

SEPTEMBER 13, 1858

In order to understand how the New England landscape was shaped by human hands, we must understand what activities its citizens pursued in the land and how they lived. Thoreau's journal is filled with everyday observations of life in town, of his neighbors, and of the countryside. These passages go far beyond the commonplace diarist's habit of noting daily activities and recording the weather in stark, simple phrases. Instead they resonate with a richness of sight, sound, and activity that convey the vibrancy of life in active New England towns where market, mill, and meadow were all part of the daily experience. As indicated by Thoreau's description above of the sound of a wheat flail, the commonplace activities of his day were actually quite transient and no longer exist in our time.

In Thoreau's descriptions we see a land brimming with activity that is strongly tied to the natural changes in the environment. Cattle were driven up-country to the hills of New England in the spring and returned in the fall. Sleighs appeared just after Thanksgiving and were then abandoned by the roadside or hidden back in the barn as the roads turned to mud in the spring. Summer brought heat and dust as well as water dripping from the ice in a deliveryman's wagon. The sawmill operated through day and night whenever the spring freshet or a periodic rainfall would allow, and it spewed its sawdust and chips into the river, where they sank and amassed in waterlogged clumps for miles along the riverbanks and pond shores. Daily routines in the house and town were equally tied to natural changes; they established a human rhythm and served as a timepiece that Thoreau could read from any vantage point in the countryside. Through the fullness

ACE OFFICE

of his descriptions, Thoreau makes the sounds, smells, feel, and sights of nineteenth-century New England seem vibrantly alive, in a way that images or facts alone could never convey. The following passages, like the pattern of change in the landscape itself, are ordered by the seasonal progression of nature and human activity.

For a week the road has been full of cattle going up country.

MAY 7, 1856

People stand at their doors in the warm evening, listening to the muttering of distant thunder and watching the forked lightning, now descending to the earth, now ascending to the clouds.

MAY 16, 1853

This is the first truly lively *summer* Sunday, what with lilacs, warm weather, waving rye, slight[ly] dusty sandy roads in some places, falling apple blossoms, etc., etc., and the wood pewee. The country people walk so quietly to church, and at five o'clock the farmer stands reading the newspaper while his cows go through the bars.

MAY 22, 1853

It is clear June [*sic*], the first day of summer. The rye, which, when I last looked, was one foot high, is now three feet high and waving and tossing its heads in the wind. We ride by these bluish-green waving rye-fields in the woods, as if an Indian juggler had made them spring up in a night. Why, the sickle and cradle will soon be taken up. Though I walk every day I am never prepared for this magical growth of the rye. I am advanced by whole months, as it were, into summer.

MAY 22, 1853

Now I see gentlemen and ladies sitting at anchor in boats on the lakes in the calm afternoons, under parasols, making use of nature.

JUNE 1, 1854

I hear now, at five o'clock, from this hill, a farmer's horn calling his hands in from the field to an early tea. Heard afar by the walker, over the woods at this hour or at noon, bursting upon the stillness of the air, putting life into some portion of the horizon, this is one of the most suggestive and pleasing of the country sounds produced by man. I know not how far it is peculiar to New England or the United States. I hear two or three prolonged blasts, as I am walking alone some sultry noon in midst of the still woods,—a sound which I know to be produced by human breath, the most sonorous parts of which alone reach me,—and I see in my mind the hired men and master dropping the implements of their labor in the field and wending their way with a sober satisfaction toward the house; I see the well-sweep rise and fall; I see the preparatory ablutions and the table laden with the smoking meal. It is a significant hum in a distant part of the hive. Often it tells me [the] time of day.

JUNE 1, 1853

Hear the sound of Barrett's sawmill, at first like a drum, then like a train of cars. The water has been raised a little by the rain after the long drought, and so he [is] obliged to saw by night, in order to finish his jobs before the sun steals it from him again.

JUNE 2, 1860

They began to carry round ice about the 1st.

JUNE 4, 1860

I see indistinctly oxen asleep in the fields, silent in majestic slumber, like the sphinx,—statuesque, Egyptian, reclining. What solid rest! How their heads are supported! A sparrow or a cricket makes more noise.

JUNE 14, 1851

I do not remember a warmer night than the last. In my attic under the roof, with all windows and doors open, there was still not a puff of the usual coolness of the night. It seemed as if heat which the roof had absorbed during the day was being reflected down upon me. It was far more intolerable than by day. All windows being open, I heard the sounds made by pigs and horses in the neighborhood and of children who were partially suffocated with the heat.

JUNE 22, 1853

As I walk through these old deserted wild orchards, half pasture, half huckleberry-field, the air is filled with fragrance from I know not what source. How much purer and sweeter it must be than the atmosphere of the streets, rendered impure by the filth about our houses! It is quite offensive often when the air is heavy at night. The roses in the front yard do not atone for the sink and pigsty and cow-yard and jakes [privies] in the rear.

JUNE 23, 1852

Very dry weather. Every traveller, horse, and cow raises a cloud of dust. It streams off from their feet, white and definite in its outline, like the steam from a locomotive. Those who walk behind a flock of sheep must suffer martyrdom. Now is that annual drought which is always spoken of as something unprecedented and out of the common course.

JULY 7, 1853

Another hot day. 96° at mid-afternoon. The elm avenue above the Wheeler farm is one of the hottest places in the town; the heat is reflected from the dusty road. The grass by the roadside begins to have a dry, hot, dusty look. The melted ice is running almost in a stream from the countryman's covered wagon, containing butter, which is to be conveyed hard to Boston market. He stands on the wheel to relieve his horses at each shelf in the ascent of Colburn Hill.

JULY 12, 1859

8.30 P.M.—The streets of the village are much more interesting to me at this hour of a summer evening than by day. Neighbors, and also farmers, come a-shopping after their day's haying, are chatting in the streets, and I hear the sound of many musical instruments and of singing from various houses.

JULY 21, 1851

This season of berrying is so far respected that the children have a vacation to pick berries, and women and children who never visit distant hills and fields and swamps on any other errand are seen making haste thither now, with half their domestic utensils in their hands. The woodchopper goes into the swamp for fuel in the winter; his wife and children for berries in the summer.

JULY 24, 1853

I hear of pickers ordered out of the huckleberry-fields, and I see stakes set up with written notices forbidding any to pick there. Some let their fields, or allow so much for the picking. *Sic transit gloria ruris.* We are not grateful enough that we have lived part of our lives before these evil days came. What becomes of the true value of country life? What if you must go to market for it?

AUGUST 6, 1858

By a gauge set in the river I can tell about what time the millers on the stream and its tributaries go to work in the morning and leave off at night, and also can distinguish the Sundays, since it is the day on which the river does not rise, but falls. If I had lost the day of the week, I could recover it by a careful examination of the river. It lies by in the various mill-ponds on Sunday and keeps the Sabbath.

AUGUST 14, 1859

The sounds heard at this hour, 8.30 [P.M.], are the distant rumbling of wagons over bridges,—a sound farthest heard of any human at night,—the baying of dogs, the lowing of cattle in distant yards.

AUGUST 15, 1845

Among other effects of the drought I forgot to mention the fine dust, which enters the house and settles everywhere and also adds to the thickness of the atmosphere.

AUGUST 27, 1854

I hear that some of the villagers were aroused from their sleep before light by the groans or bellowings of a bullock which an unskillful butcher was slaughtering at the slaughter-house. What morning or Memnonian music was that to ring through the quiet village? What did that clarion sing of? What a comment on our village life! Song of the dying bullock! But no doubt those who heard it inquired, as usual, of the butcher the next day, "What have you got to-day?" "Sirloin, good beefsteak, rattleran," etc.

AUGUST 28, 1859

Cattle are driven down from up-country. Hear the drovers' *whoa whoa whoa* or *whay whay whay.*

AUGUST 29, 1854

1 A.M., moon waning, to Conantum. A warm night. A thin coat sufficient. I hear an apple fall, as I go along the road. Meet a man going to market thus early.

SEPTEMBER 3, 1852

Bathed at the swamp white oak, the water again warmer than I expected. One of these larger oaks is stripped nearly bare by the caterpillars. Cranberry-raking is now fairly begun. The very bottom of the river there is loose and crumbly with sawdust. I bring up the coarse *bits* of wood (water-logged) between my feet.

SEPTEMBER 5, 1854

Coming home through the street in a thunder-shower at ten o'clock this night, it was exceedingly dark. I met two persons within a mile, and they were obliged to call out from a rod distant lest we should run against each other.

SEPTEMBER 18, 1857

Stopped at Barrett's mill. He had a buttonwood log to saw. In an old grist-mill the festoons of cobwebs revealed by the white dust on them are an ornament. Looking over the shoulder of the miller, I drew his attention to a mouse running up a brace. "Oh, yes," said he, "we have plenty of them. Many are brought to the mill in barrels of corn, and when the barrel is placed on the platform of the hopper they scamper away."

SEPTEMBER 25, 1857

All sorts of men come to Cattle-Show. I see one with a blue hat.

SEPTEMBER 29, 1857

5 P.M.—Just put a fugitive slave, who has taken the name of Henry Williams, into the cars for Canada. He escaped from Stafford County, Virginia, to Boston last October; has been in Shadrach's place at the Cornhill Coffee-House; had been corresponding through an agent with his master, who is his father, about buying himself, his master asking $600, but he having been able to raise only $500. Heard that there were writs out for two Williamses, fugitives, and was informed by his fellow-servants and employer that Augerhole Burns and others of the police had called for him when he was out. Accordingly fled to Concord last night on foot, bringing a letter to our family from Mr. Lovejoy of Cambridge and another which [William Lloyd] Garrison had formerly given him on another occasion. He lodged with us, and waited in the house till funds were collected with which to forward him. Intended to dispatch him at noon through to Burlington, but when I went to buy his ticket, saw one at

the depot who looked and behaved so much like a Boston policeman that I did not venture that time.

OCTOBER 1, 1851

One shopkeeper has hung out woollen gloves and even thick buckskin mittens by his door, foreseeing what his customers will want as soon as it is finger-cold, and determined to get the start of his fellows.

OCTOBER 26, 1858

I heard one boy say to another in the street to-day, "You don't know much more than a piece of putty."

OCTOBER 28, 1852

To-night a free colored woman is lodging at our house, whose errand to the North is to get money to buy her husband, who is a slave to one Moore in Norfolk, Virginia. She persuaded Moore, though not a kind master, to buy him that he might not be sold further South.

NOVEMBER 1, 1853

This evening at sundown, when I was on the water, I heard come booming up the river what I suppose was the sound of cannon fired in Lowell to celebrate the Whig victory, the voting down the new Constitution.

NOVEMBER 15, 1853

As we were walking through West Acton the other afternoon, a few rods only west of the centre, on the main road, the Harvard turnpike, we saw a rock larger than a man could lift, lying in the road, exactly in the wheel-track, and were puzzled to tell how it came there, but supposed it had slipped off a drag,—yet we noticed that it was peculiarly black. Returning the same way in the twilight, when we had got within four or five rods of this very spot, looking up, we saw a man in the field, three or four rods on one side of that spot, running off as fast as he could. By the time he had got out of sight over the hill it occurred to us that he was blasting rocks and had just touched one off; so, at the eleventh hour, we turned about and ran the other way, and when we had gone a few rods, off went two blasts, but fortunately none of the rocks struck us. Some time after we had passed we saw the men returning. They looked out for themselves, but for nobody else. This is the way they do things in West Acton.

NOVEMBER 17, 1860

Already you see the tracks of sleds leading by unusual routes, where will be seen no trace of them in summer, into far fields and woods, crowding aside and pressing down the snow to where some heavy log or stone has thought itself secure, and the spreading tracks also of the heavy, slow-paced oxen, of the well-shod farmer, who turns out his feet. Ere long, when the cold is stronger, these tracks will lead the walker deep into remote swamps impassable in summer. All the earth is a highway then.

DECEMBER 4, 1856

I stood by Bigelow the blacksmith's forge yesterday, and saw him repair an axe. He burned the handle out, then, with a chisel, cut off the red-hot edge even, there being some great gaps in it, and by hammering drew it out and shaped it anew,—all in a few minutes. It was interesting to see performed so simply and easily, by the aid of fire and a few rude tools, a work which would have surpassed the skill of a tribe of savages.

DECEMBER 14, 1855

Boys are now devoted to skating after school at night, far into evening, going without their suppers.

DECEMBER 20, 1855

After talking with Uncle Charles the other night about the worthies of this country, [Daniel] Webster and the rest, as usual, considering who were geniuses and who not, I showed him up to bed, and when I had got into bed myself, I heard his chamber door opened, after eleven o'clock, and he called out, in an earnest, stentorian voice, loud enough to wake the whole house, "Henry! was John Quincy Adams a genius?" "No, I think not," was my reply. "Well, I didn't think he was," answered he.

JANUARY 1, 1853

Mother tried to milk the cow which Father took on trial, but she kicked at her and spilt the milk. (They say a dog had bitten her teats.) Proctor laughed at her as a city girl, and then he tried, but the cow kicked him over, and he finished by beating her with his cowhide shoe. Captain Richardson milked her warily, standing up. Father came home, and thought he would "brustle right up to her," for she needed much to be milked, but suddenly she lifted her leg and "struck him fair and square right in the muns," knocked him flat, and broke the bridge of his nose, which shows it yet. He distinctly heard her hoof rattle on his nose. This "started the claret," and, without stanching the blood, he at once drove her home to the man he had her of. She ran at some young women by the way, who saved themselves by getting over the wall in haste.

JANUARY 7, 1856

The milkman is now filling his ice-house.

JANUARY 22, 1852

To get ice at all clear or transparent, you must scrape the snow off after each fall. Very little ice is formed by addition below, such a snowy winter as this.

JANUARY 23, 1856

How simple the machinery of the mill! Miles [the miller] has dammed a stream, raised a pond or head of water, and placed an old horizontal mill-wheel in position to receive a jet of water on its buckets, transferred the motion to a horizontal shaft and saw by a few cog-wheels and simple gearing, and, throwing a roof of slaps over all, at the outlet of the pond, you have a mill.

FEBRUARY 28, 1856

Stopped at Martial Miles's to taste his cider. Marvellously sweet and spirited without being bottled; alum and mustard put into the barrels.

FEBRUARY 28, 1856

A lady tells me that she met Deacon S. of Lincoln with a load of hay, and she, noticing that as he drove under the apple trees by the side of the road a considerable part of the hay was raked off by their boughs, informed him of it. But he answered, "It is not mine yet. I am going to the scales with it and intend to come back this way."

MARCH 8, 1861

When March arrives, a tolerably calm, clear, sunny, spring-like day, the snow is so far gone that sleighing ends and our compassion is excited by the sight of horses laboriously dragging wheeled vehicles through mud and water and slosh. We shall no longer hear the jingling of sleighbells. The sleigh is housed, or, perchance, converted into a wheeled vehicle by the travelling peddler caught far from home. The wood-sled is perhaps abandoned by the roadside, where the snow ended, with two sticks put under its runners,—there to rest, it may be, while the grass springs up green around it, till another winter comes round. It may be near where the wagon of the careless farmer was left last December on account of the drifted snow. As March approaches, at least, peddlers will do well to travel with wheels slung under their sleighs, ready to convert their sleighs into wheeled vehicles at an hour's warning.

MARCH 25, 1860

The boy's sled gets put away in the barn or shed or garret, and there lies dormant all summer, like a woodchuck in the winter. It goes into its burrow just before woodchucks come out, so that you may say a woodchuck never sees a sled, nor a sled a woodchuck,—unless it were a prematurely risen woodchuck or a belated and unseasonable sled. Before the woodchuck comes out the sled goes in. They dwell at the antipodes of each other. Before sleds rise woodchucks have set. The ground squirrel too shares the privileges and misfortunes of the woodchuck. The sun now passes from the constellation of the sled into that of the woodchuck.

MARCH 25, 1860

The Farmer as Hero

But how much clearing of land and plowing and planting and building of stone wall is done every summer without being reported in the newspapers or in literature! Agricultural literature is not as extensive as the fields, and the farmer's almanac is never a big book. And yet I think that the history (or poetry) of one farm from a state of nature to the highest state of cultivation comes nearer to being the true subject of a modern epic than the siege of Jerusalem or any such paltry and ridiculous resource to which some have thought men reduced.

MARCH 2, 1852

Agriculture was the main force driving the phenomenal episode of forest clearance that created the widely open landscape of Henry Thoreau's day. Most of the forest land that we enjoy today occupies that former agrarian expanse. Consequently, if we are to understand the historical transformation of New England and evaluate the factors influencing modern forest landscapes, we need to understand nineteenth-century agricultural practices and to assess their impact on the land. Specifically, it is important first to know how farming practices altered the existing natural features of the land and then to determine the extent to which these changes have persisted through time and are left as legacies of the past. In pursuit of answers to these questions, Thoreau's observations are extremely revealing.

Thoreau recognized and thoroughly appreciated the staggering amount of work that was routinely undertaken by farmers as they pursued their daily activities that transformed and maintained the land. From his perspective as he wandered the fields and laneways, the tasks associated with agriculture were Herculean, and the epic proportions of these efforts made the farmer a heroic figure. Although agriculture was generally dismissed as a common and routine activity by many of his contemporaries, Thoreau felt that, if it were assessed in broad perspective, the work of the farmer would warrant major attention and acclaim. Since New England farmers work continually, in a seasonally changing climate, their activities fit a regular and quite predictable annual cycle that Thoreau documented in considerable detail. Spring plowing and planting led to a sequence of summer harvests, while the late fall and winter were marked by maintenance activities—land clearance, fuelwood cutting, brush burning, stone wall building, ditch digging and cleaning, and manuring. It is the fine details provided by Thoreau's journal within this seasonal pattern that remind us of our modern distance from historical land-use practices and that provide a wealth of information on the immediate impacts and enduring consequences of these activities.

Nineteenth-century agricultural New England contrasted strongly with the image that modern scenes of dairy farms, with their red barns and white houses, present to the traveler today. Farms in Thoreau's day produced the diverse crops that met most of the local and regional needs, including vegetables and grains, meats, dairy products, and eggs. In an era of ploddingly slow land travel and limited refrigeration, milk was produced locally, while beef, which could be transported rather easily on the hoof, provided a major export product for many farm communities across New England. Consequently, the beef trade and related agricultural activities created a distinctive annual pattern in the countryside. Cows were fed hay and grain in barns through the winter and then in mid-spring were driven to the up-country pastures and hill towns of Massachusetts, New Hampshire, and Vermont, where they grazed, calved, and were tended by herdsmen and young boys. In the fall they were driven down to home pastures or east to Brighton, the large butchering yard and cattle market on the outskirts of Boston. Once butchered, their hides were tanned for leather, and the meat was salted, packed, and shipped down the coast and through the Caribbean to the West Indies, thereby forming a leg in the triangular trade route that connected the American and Caribbean tropics with the temperate Old and New Worlds. This seasonal cycle of beef farming had an indirect impact on Thoreau's walks, for he contrasted the bustle of people and animals moving across the countryside through the growing season with the "unnatural" solitude of the fields and hills that prevailed during the winter.

In order to feed both cattle and people, fields were plowed, harrowed, and planted with an abundance of corn and many grains that we associate today with the broad plains of the American Midwest, including oats, rye, barley, and wheat. Thoreau remarks frequently on the density, height, and extent of these fields of grain. Their color and pattern created an appearance that would be quite foreign to any modern New Englander but one strikingly familiar to travelers in Britain or Scandinavia today, where the small size and limited range of the landscape bring temperate forest and wheat fields into close juxtaposition. Associated with the grains and the widespread fields of hay were all the important implements of land use: the sickle, scythe, cylinder, harrow, and flail.

Thoreau also notes the extensive efforts of farmers and their hired hands in the ditching and manuring of fields and related activities that increased the size and improved the drainage and productivity of arable lands. Swamps, bogs, and lowlands were cleared of trees and shrubs and then interconnected by networks of ditches in an effort to dry and aerate their saturated soils. These ditches, which were nearly level, were forever clogging and collapsing, and they required annual cleaning and maintenance in order to maximize their effectiveness. However, their long-term impact was to reduce permanently the vast areas of wetlands and wet soils across the land. In extreme cases, the swamps and marshes were dug and scraped after the forest was cleared, and the resulting stumps, peat, and muck were piled and burned when dry.

The array of manures used in farming and the close attention that farmers paid to improving the soil form an interesting element in Thoreau's observations because they record an awareness of the natural limitations in fertility of the thin, rocky, and mostly acidic New England soils. Given the extent of agricultural activity in nineteenth-century New England, these observations also suggest that farming practices undoubt-

edly initiated many long-term changes in soil chemistry and structure that continue to affect the forests that now occupy these areas. Manure from the barns was carted to the fields; muck from lake shores and swamps was piled to dry and then spread during plowing; and farmers gathered clamshells, ashes, potash, and various other materials as sources of lime and potassium. The agricultural necessity of getting the best use out of every possible source of fertilizer is highlighted by Thoreau's repeated observations of farmers traversing their fields knocking apart piles of manure to even its distribution and increase its effectiveness.

Along with the hydrological and chemical changes in the soil that were initiated by farming activity, Thoreau noted many physical impacts as well. Stones were removed and placed in walls to improve the land and make plowing easier. Cultivation of the soil after fall harvest was a routine practice followed by many farmers that left the surface bare and exposed to wind, rain, and the action of freeze and thaw throughout the winter. On hill slopes, the spring rains could easily erode these extensive uncovered areas, moving finer soils to the bottom of the slopes where they filled depressions or accumulated against the stone walls that marked the end of the field. In an interesting commentary on the extreme environmental consequences of agriculture, Thoreau suggests in his journal that even the best efforts to counteract erosion would fail on the steepest slopes and that many such sites would better be left as woodland. The extent to which humans and their animals transformed the landscape was profound, inescapable, and enduring.

In balance, and perhaps counter to the sentiment that emerges from the pages of *Walden,* Thoreau viewed much of farmers' activity as noble and heroic; it provided a solid and physical counterbalance to his own mental activities as he traversed the land. His descriptions capture a different age and highlight many agricultural practices that may be quite distinct from modern New Englanders' view of their land's history. These nineteenth-century farmers did more than clear forest and raise crops; in many ways, they altered the quality and type of land permanently. Thus these earlier practices still have a profound impact on the landscape and vegetation of New England.

> After having read various books on various subjects for some months, I take up a report on Farms by a committee of Middlesex Husbandmen, and read of the number of acres of bog that some farmer has redeemed [i.e., made into productive agricultural land], and the

number of rods of stone wall that he has built, and the number of tons of hay he now cuts, or of bushels of corn or potatoes he raises there, and I feel as if I had got my foot down on to the solid and sunny earth, the basis of all philosophy, and poetry, and religion even. I have faith that the man who redeemed some acres of land the past summer redeemed also some parts of his character. I shall not expect to find him ever in the almshouse or the prison. He is, in fact, so far on his way to heaven. When he took the farm there was not a grafted tree on it, and now he realizes something handsome from the sale of fruit. These, in the absence of other facts, are evidence of a certain moral worth.

MARCH 1, 1852

The farmer increases the extent of the habitable earth. He makes soil. That is an honorable occupation.

MARCH 2, 1852

The race that settles and clears the land has got to deal with every tree in the forest in succession. It must be resolute and industrious, and even the stumps must be got out,—or are. It is a thorough process, this war with the wilderness,—breaking nature, taming the soil, feeding it on oats. The civilized man regards the pine tree as his enemy. He will fell it and let in the light, grub it up and raise wheat or rye there. It is no better than a fungus to him.

FEBRUARY 2, 1852

This new pasture, with gray stumps standing thickly in the now sere sward, reminds me of a graveyard. And on these monuments you can read each tree's name, when it was born (if you know when it died), how it throve, and how long it lived, whether it was cut down in full vigor or after the infirmities of age had attacked it.

FEBRUARY 7, 1858

I saw an improvement, I suppose by William Brown, on the shore of the pond this afternoon, which really is something to tell of. The exploits of the farmer are not often reported even in the agricultural paper, nor are they handed down by tradition from father to son, praiseworthy and memorable as so many of them are; though if he ran away from hard work once in his youth and enlisted, and chanced to be present at one short battle, he will even in his old age love to dwell on this, "shoulder his crutch and *show how fields are won,*" with cruel satire, as if he had not far better shown this with his axe and spade and plow. Here was an extensive swamp, level of course as a floor, which first had been cut, then ditched broadly, then burnt over; then the surface paved off, stumps and all, in great slices; then these piled up every six feet, three or four feet high, like countless larger muskrat-cabins, to dry; then fire put to them; and so the soil was tamed. We witnessed the different stages in different parts of the swamp.

JANUARY 28, 1853

Shannon tells me that he took a piece of bog land of Augustus Hayden, cleared, turned up the stumps and roots and burned it over, making a coat of ashes six inches deep, then planted potatoes. He never put a hoe to it till he went to dig them; then between 8 o'clock A.M. and 5 P.M. he and another man dug and housed seventy-five bushels apiece!!

AUGUST 31, 1851

I have no doubt that a good farmer, who, of course, loves his work, takes exactly the same kind of pleasure in draining a swamp, seeing the water flow out in his newly cut ditch, that a child does in its mud dikes and water-wheels. Both alike love to play with the natural forces.

NOVEMBER 8, 1857

Cows spend their winters in barns and cow-yards, their summers in pastures. In summer, therefore, they may low with emphasis, "To-morrow to fresh woods and pastures new." I sometimes see a neighbor or two united with their boys and hired men to drive their cattle to some far-off country pasture, fifty or sixty miles distant in New Hampshire, early in the morning, with their sticks and dogs. It is a memorable time with the farmers' boys, and frequently their first journey from home. The herdsman in some mountain pasture is expecting them. And then in the fall, when they go up to drive them back, they speculate as to whether Janet or Brindle will know them. I heard such a boy exclaim on such an occasion, when the calf of the spring returned a heifer, as he stroked her side, "She knows me, father; she knows me." Driven up to be the cattle on a thousand hills.

MAY 31, 1850

Cattle begin to go up-country, and every week day, especially Mondays, to this time [*sic*] May 7th, at least, the greatest droves to-day. Methinks they will find slender picking up there for a while. Now many a farmer's boy makes his first journey, and sees something to tell of,—makes acquaintance with those hills which are mere blue warts in his horizon, finds them solid and *terra firma,* after all, and inhabited by herdsmen, partially befenced and measurable by the acre, with cool springs where you may quench your thirst after a dusty day's walk.

APRIL 30, 1860

Yesterday I observed many fields newly plowed, the yellow soil looking very warm and dry in the sun; and one boy had fixed his handkerchief on a stick and elevated it on the yoke, where it flapped or streamed and rippled gayly in the wind, as he drove his oxen dragging a harrow over the plowed field.

APRIL 30, 1852

The farmers are very busily harrowing and rolling in their grain. The dust flies from their harrows across the field. The tearing, toothed harrow and the ponderous cylinder, which goes creaking and rumbling over the surface, heard afar, and vying with the sphere.

MAY 6, 1852

It rains gently from time to time as I walk, but I see a farmer with his boys, John Hosmer, still working in the rain, bent on finishing his planting. He is slowly getting a soaking, quietly dropping manure in the furrows.

MAY 17, 1858

As I come out of the Spring Woods I see Abiel Wheeler planting peas and covering them up on his warm sandy hillside, in the hollow next the woods. It is a novel sight, that of the farmer distributing manure with a shovel in the fields and planting again.

MARCH 26, 1857

R. W. E. [Ralph Waldo Emerson] tells me he does not like Haynes as well as I do. I tell him that he makes better manure than most men.

MAY 4, 1852

Indeed the farmer's was pretty much the same routine then [in Classical times] as now. Cato [Marcus Porcius Cato, 234–149 B.C., a Roman statesman, orator, and writer whose treatise on agriculture is the oldest complete work in Latin] says: "Sterquilinium magnum stude ut habeas. Stercus sedulo conserva, cum exportabis purgato et comminuito. Per autumnum evehito." (Study to have a great dungheap. Carefully preserve your dung, when you carry it out, make clean work of it and break it up fine. Carry it out during the autumn.) Just such directions as you find in the "Farmer's Almanack" to-day. It reminds me of what I see going on in our fields every autumn. As if the farmers of Concord were obeying Cato's directions. And Cato but repeated the maxims of a remote antiquity.

SEPTEMBER 2, 1851

Some of these shells at Clamshell Hill, whose contents were cooked by the Indians, are still entire, but separated. Wood has spread a great many loads over his land.

DECEMBER 3, 1853

The other day I saw what I took to be a scarecrow in a cultivated field, and noticing how unnaturally it was stuffed out here and there and how ungainly its arms and legs were, I thought to myself, "Well, it is thus they make these things; they do not stand much about it;" but looking round again after I had gone by, I saw my scarecrow walking off with a real live man in it.

JUNE 23, 1853

There are certain crops which give me the idea of bounty, of the *Alma Natura.* They are the grains. Potatoes do not so fill the lap of earth. This rye excludes everything else and takes possession of the soil. The farmer says, "Next year I will raise a crop of rye;" and he proceeds to clear away the brush, and either plows it, or, if it is too uneven or stony, burns and harrows it only, and scatters the seed with faith. And all winter the earth keeps his secret,—unless it did leak out somewhat in the fall,—and in the spring this early green on the hillsides betrays him.

JULY 8, 1851

Here are some rich rye-fields waving over all the land, their heads nodding in the evening breeze with an apparently alternating motion; *i. e.* they do not all bend at once by ranks, but separately, and hence this agreeable alternation. How rich a sight this cereal fruit, now yellow for the cradle,—*flavus!* It is an impenetrable phalanx. I walk for half a mile beside these Macedonians, looking in vain for an opening.

JULY 8, 1851

Haying is fairly begun, and for some days I have heard the sound of the mowing-machine, and now the lark must look out for the mowers.

JULY 11, 1857

How handsome lie the oats which have been cradled in long rows in the field, a quarter of a mile uninterruptedly! The thick stub ends, so evenly laid, are almost as rich a sight to me as the graceful tops.

AUGUST 2, 1854

I see where a field of oats has been cradled, by the railroad,—alternate white and dark green stripes, the width of a swath, running across the field. I find it arises from the stubble being bent a particular way by the cradle, as the cradler advanced, and accordingly reflecting the light but one way, and if I look over the field from the other side, the first swath will be dark and the latter white.

OCTOBER 12, 1851

For six weeks or more this has been the farmer's work, to shave the surface of the fields and meadows clean. This is done all over the country. The razor is passed over these parts of nature's face the country over. A thirteenth labor which methinks would have broken the back of Hercules, would have given him a memorable sweat, accomplished with what sweating of scythes and early and late! I chance [to] know one young man who has lost his life in this season's campaign, by overdoing. In haying time some men take double wages, and they are engaged long before in the spring. To shave all the fields and meadows of New England clean! If men did this but once, and not every year, we should never hear the last of that labor; it would be more famous in each farmer's case than Buonaparte's road over the Simplon. It has no other bulletin but the truthful "Farmer's Almanac." Ask them where scythe-snaths are made and sold, and rifles too, if it is not a real labor. In its very weapons and its passes it has the semblance of war. Mexico was won with less exertion and less true valor than are required to do one season's haying in New England. The former work was done by those who played truant and ran away from the latter. Those Mexicans were mown down more easily than the summer's crop of grass in many a farmer's fields. Is there not some work in New England men? This haying is no work for marines, nor for deserters; nor for United States troops, so called nor for West Point cadets. It would wilt them, and they would desert. Have they not deserted? and run off to West Point?

AUGUST 17, 1851

Hear the sound of the first flail,—some farmer, perchance, wishing to make room in his barn, or else wanting the grain. Is it wheat or rye? It may be either.

JULY 30, 1860

Early for several mornings I have heard the sound of a flail.

AUGUST 29, 1854

The farmers now carry—those who have got them—their early potatoes and onions to market, starting away early in the morning or at midnight. I see them returning in the afternoon with the empty barrels.

AUGUST 23, 1851

I hear A—— W—— complained of for overworking his cattle and hired men, but there is this to be said in his favor, that he does not spare himself. They say that he made his horse "Tom" draw twenty-nine hundred of hay to Boston the other day,—or night,—but then he put his shoulder to the wheel at every hill. I hear that since then the horse has died, but W—— is alive and working.

AUGUST 29, 1858

People now (at this low stage of water) dig mud for their compost-heaps, deepen wells, build bank walls, perchance, along the river, and in some places make bathing-places by raking away the weeds. Many are ditching.

AUGUST 21, 1859

The farmers improve this season, which is the driest, their haying being done and their harvest not begun, to do these jobs,—burn brush, build walls, dig ditches, cut turf. This is what I find them doing all over the country now; also topping corn and digging potatoes.

SEPTEMBER 4, 1851

His [Sam Barrett's] pond has been almost completely dry,—more than he ever knew,—and is still mostly so. The muddy bottom is exposed high and dry, half a dozen rods wide, and half covered with great drying yellow and white lily pads and stems. He improves the opportunity to skim off the fertile deposit for his compost-heap.

SEPTEMBER 5, 1854

Here comes a laborer from his dinner to resume his work at clearing out a ditch notwithstanding the rain, remembering as Cato says, *per ferias potuisse fossas veteres tergeri,* that in the holidays old ditches might have been cleared out.

SEPTEMBER 3, 1851

See the heaps of apples in the fields and at the cider-mill, of pumpkins in the fields, and the stacks of cornstalks and the standing corn. Such is the season.

OCTOBER 10, 1857

This is the seventh day of glorious weather. Perhaps these might be called Harvest Days. Within the week most of the apples have been gathered; potatoes are being dug; corn is still left in the fields, though the stalks are being carried in. Others are ditching and getting out mud and cutting up bushes along fences,—what is called "brushing up,"—burning brush, etc.

OCTOBER 11, 1857

I saw a farmer busily collecting his pumpkins on the 14th,—Abel Brooks,—rambling over his corn-fields and bringing the pumpkins out to the sides on the path, on the side of the field, where he can load them. The ground was so stiff on the 15th, in the morning, that some could not dig potatoes. Bent is now making haste to gather his apples. I. Wright, too, is collecting some choice barrels of golden russets. Many times he turns it over before he leaves out a specked one.

OCTOBER 16, 1856

The farmers are now casting out their manure, and removing the muck-heap from the shore of ponds where it will be inaccessible in the winter; or are doing their fall plowing, which destroys many insects and mellows the soil. I also see some pulling their turnips, and even getting in corn which has been left out notwithstanding the crows. Those who have wood to sell, as the weather grows colder and people can better appreciate the value of fuel, lot off their woods and advertise a wood auction.

NOVEMBER 15, 1850

I rode home from the woods in a hay-rigging, with a boy who had been collecting a load of dry leaves for the hog-pen; this the third or fourth load.

DECEMBER 5, 1853

It snowed in the night of the 6th, and the ground is now covered,—our first snow, two inches deep. A week ago I saw cows being driven home from pasture. Now they are kept at home. Here's an end to their grazing. The farmer improves this first slight snow to accomplish some pressing jobs,—to move some particular rocks on a drag, or the like. I perceive how quickly he has seized the opportunity. I see no tracks now of cows or men or boys beyond the edge of the wood. Suddenly they are shut up. The remote pastures and hills beyond the woods are now closed to cows and cowherds, aye, and to cowards. I am struck by this sudden solitude and remoteness which these places have acquired.

DECEMBER 8, 1850

At Willow Bay I see for many rods black soil a quarter of an inch deep, covering and concealing the ice (for several rods). This, I find, was blown some time ago from a plowed field twenty or more rods distant. This shows how much the sediment of the river may be increased by dust blown into it from the neighboring fields.

JANUARY 26, 1859

Walking afterward on the side of the hill behind Abel Hosmer's, overlooking the russet interval, the ground being bare where corn was cultivated last year, I see that the sandy soil has been washed far down the hill for its whole length by the recent rains combined with the melting snow, and it forms on the nearly level ground at the base very distinct flat yellow sands, with a convex edge, contrasting with the darker soil there. Such slopes must lose a great deal of this soil in a single spring, and I should think that was a sound reason in many cases for leaving them woodland and never exposing and breaking the surface. This, plainly, is one reason why the brows of such hills are commonly so barren. They lose much more than they gain annually. It is a question whether the farmer will not lose more by the wash in such cases than he will gain by manuring.

MARCH 19, 1859

It is interesting to see loads of hay coming down from the country nowadays,—within a week. They make them very broad and low. They do not carry hay by railroad yet. The spoils of up-country fields. A mountain of dried herbs. I had forgotten that there ever was so much grass as they prove.

FEBRUARY 8, 1852

Now and for some days I see farmers walking about their fields, knocking to pieces and distributing the cow-dung left there in the fall, that so, with the aid of the spring rains, they fertilize a larger surface and more equally.

MARCH 8, 1860

These slight falls of snow which come and go again so soon when the ground is partly open in the spring, perhaps helping to open and crumble and prepare it for the seed, are called "the poor man's manure." They are, no doubt, more serviceable still to those who are rich enough to have some manure spread on their grass ground, which the melting snow helps dissolve and soak in and carry to the roots of the grass. At any rate, it is all the poor man has got, whether it is good or bad.

MARCH 18, 1852

The maple sap has been flowing well for two or three weeks.

MARCH 28, 1857

The yellow birch sap runs very fast. I set three spouts in a tree one foot in diameter, and hung on a quart pail; then went to look at the golden saxifrage in Hubbard's Close. When I came back, the pail was running over.

APRIL 10, 1856

Meadows and Mowers

The lark sings at sundown off in the meadow. It is a note which belongs to a New England summer evening.

JUNE 30, 1851

The word "meadow" brings to mind images of native grasses and a mixture of multicolored wildflowers, fields with straggling wild roses and strawberries in more open places, and bobolinks, meadowlarks, and red-winged blackbirds twittering from arching stems. It's a wonderful image, but how many of us have actually walked through one? Or know where to find one? Once an extremely common part of the New England landscape, meadows have disappeared at a phenomenal rate in this century. Today in the eastern United States, as in Britain and elsewhere across the European continent, meadows are rare but highly desirable for the sights, wildlife, and habitat that they offer. Landowners expend a great deal of effort and money hiring farmers or landscape gardeners to create or maintain these and other open grasslands and shrublands. Conservation organizations, including the Audubon Society and English Nature (The Nature Conservancy of England) have identified meadows and other semi-natural grasslands as critical habitats for a rapidly decreasing group of insects and other invertebrates, birds, small mammals, and plants. Meadows are now threatened and have become an environmental priority.

What process allowed these grassy areas to exist in the middle of a landscape that was naturally dominated by forest, and why have they shrunk in abundance? In New England in the time of the colonists, their agrarian descendants, and Henry Thoreau, the term "meadow" had a specific and quite restricted meaning, referring to the grassy and uncultivated lowland areas bordering rivers and similar sites of low, moist vegetation. Fields on the uplands might consist of cultivated grains, scraggly pastures of chewed and matted grasses, thistles and shrubs, or "mowings" covered with imported English grasses. On true lowland meadows, the predominant grasses and sedges were native, and the lack of cultivation encouraged a diverse assortment of herbs, ferns, and woody plants to grow in their midst.

For Thoreau, who spent many a day boating or walking the banks of the Sudbury, Assabet, and Concord Rivers, meadows were a dominant part of the landscape, and their natural history, beauty, and human activity captured his attention and were often described in his journal. With its broad and flat lowlands and sluggish, meandering rivers, Concord was better endowed with marshes and meadows than many New England towns. Nonetheless, most town histories and tax accounts from across the region contain descriptions of meadows, and their importance to the first settlers and to the eighteenth- and nineteenth-century agricultural economy is hard to overstate. As Thoreau saw them, the meandering and slowly flowing rivers were really chains of flooded meadows that underwent a predictable and captivating annual cycle. As the snow melted in late winter, the spring freshets swelled the streams and flooded the adjoining grassy areas, enabling adventurers like him to boat across them; these freshets deposited river debris, silt, and organic material and scoured the land and river banks with ice. Through the summer months as the high water subsided, the grasses and a wide array of other plants grew to bend in the wind and provided cover and food to distinctive birds—snipe, plover, peeps, yellowlegs, bittern, heron, meadowlark, and bobolink—as well as muskrat, turtles, frogs, and a variety of colorful butterflies. In the weeks after the upland hay fields were mown, the farmers descended in droves to cut the moister meadows and to deliver the grass safely to barns before a sudden rain, or the negligence of a damkeeper or mill operator, led to a rising of the river and flooding of the newly cut hay. Thus, Thoreau enjoyed the meadows for their distinctive mixture of nature and human presence.

The annual meadow cycle that Thoreau witnessed in Concord was age-old and thoroughly established across New England. From the earliest days of the colonists, grassy wetlands, both freshwater marshes along rivers and salt marshes at the coast, provided an essential source of grass and forage for farm animals. As the first forest areas were being cleared to make way for upland hay fields and pasture, the "natural" lowland meadows provided grazing land for the hungry animals, and stored meadowgrass fed them through the harsh winter months. The colonists rapidly learned a number of ways to augment the natural production of the meadows and improve the annual harvest by favoring grasses over the shrubs and bushes that sought to crowd them out. These included regulating the water table by ditching the lowlands and damming the streams, cutting the meadow grasses and herbs low and thoroughly each year, and setting an occasional fire to eliminate the woody plants.

Thus the meadows, along with their adjoining streams, played a role in the seasonal economy of New England farms and provided an important source of waterfowl, fish, and fur-bearing animals, especially muskrat. The natural timing of slow growth and late maturation of the native lowland grasses enabled them to be harvested later than the imported upland varieties. Consequently, even after upland areas across New England were well developed as productive agricultural land, the lowland grasses continued to provide a significant amount of hay. However, as Thoreau writes, the general consensus among his farmer acquaintances was that meadow hay was coarser and slightly inferior to the imported grasses on the cultivated fields. Even so, he delights in noting that the cattle received a more splendid feed from the meadow—it provided fodder worthy of botanizing, with its diversity of plants. The hay from some meadows attracted certain cows with its prevalence of pipes (horsetails or *Equisetum* species) and others with its ferns, and always yielded a mixture of grasses, sedges, wildflowers, and young shoots of woody plants. In fact, to Thoreau, this use of the meadows' diverse native plants by farmers provided a historical connection from his landscape back to the longstanding European and early colonial tradition of feeding livestock on leaves and twigs.

From the extent and vividness of his journal passages on the subject, it is clear that Thoreau delighted in the human element of the meadows and recognized it as an integral part of his landscape: the phalanxes of mowers in staggered, diagonal rows; the boys piling and windrowing the cut grass with long-handled rakes; the horses pulling large mechanical rakes; the scattered haycocks and fully loaded wagons; the men and animals resting in the shade of a hot midsummer's day; and the jugs of water, molasses, and blackstrap liquor passed among them. Like upland hay, the meadow grass needed to be cut, dried in the hot sun and wind, raked into loose rows or piles for further drying, and then bundled into haycocks and loaded onto wagons. However, in contrast to the uplands, where the sole concern for the farmer was that a sudden rain might interrupt the drying process and yield wet, moldy, or musty hay, in the meadows the distant action of a self-centered or cantankerous dam owner bent on altering the water flow for his own purposes could generate a flood that would inundate, mildew, and quickly blacken the hay. Thus haying always brought a time of hard work, sweat, waiting, drinking, and worrying. Unlike the

uplands, where the fields were an integral part of one's farm and were well marked by stone walls and fences, the lowland meadows were often located quite a distance from the main farm and, lacking permanent boundaries, were less obviously demarcated. In many towns, the meadows were divided up into narrow strips so that citizens could own and share in the harvest of this important crop. These elongated lots extended from the upland border to the river's edge, and before each mowing they were marked off by sticks adorned with newspaper or leaves. These were the details that Thoreau the surveyor noted: the owners cutting narrow swaths along extended rectangular blocks, mowing side by side, guided by the makeshift boundary markers.

In his musings on the subject of meadows, Thoreau frequently returned to some central questions of their natural history that are important in understanding the changes that have occurred in our own landscape. What actually maintained the openness of these grassy areas? Where are the natural meadows located, if any exist? What were meadows like in the past before the intensive mowing and grazing began—wetter or drier, larger or fewer? The hordes of small saplings that Thoreau saw crowding out from the upland margin of the meadows suggested that there was, in fact, a tendency for trees and shrubs to invade across these areas and smother the grasses with their growth. He also became convinced, through his own observations and those of his farmer acquaintances, that the mowers' tendency toward laziness was allowing bushes and trees to establish themselves with increasing frequency. Clearly, he noted, cutting, burning, and grazing by livestock were important in maintaining the dominant growth of grass, sedge, ferns, and flowers. And yet the death of trees that occurred with an episode of unusually high flooding, particularly in summer after the plants had leafed out, indicated that the annual inundation of the river across the soft and muddy ground was deleterious to the growth and spread of trees. Thus there did appear to be a natural factor involved in the formation and persistence of the meadows. Apparently, Thoreau concluded after years of examination and speculation, human activity increases a natural tendency toward grass and herbs on these lowland sites; without this activity, the open grasslands would be qualitatively different and quantitatively less abundant.

Thoreau's observations contain a strong message that is beginning to gain recognition among conservationists and landowners interested in protecting and maintaining meadows as well as other grasslands and shrublands in New England. The human activity that Thoreau witnessed—the ditching, burning, grazing, water regulation, and, above all, the repeated annual cutting—was essential for the creation and persistence of most meadows and other open areas that the countryside supported in his day. These practices were also necessary to maintain the diversity and array of plants and animals that he marveled at in the course of his boat trips and walks. Over the years, as the intense activity associated with meadow haying declined with the regional abandonment of farms and farming, the woodiness of the meadows increased. Consequently, by discontinuing the nineteenth-century methods of land management and by controlling the natural extremes of flooding, we have lost most of the meadows and many other elements of the historical landscape, along with much of the rural character of New England that we have read about, but cannot locate around us today. In order to retain such features of the classic New England landscape we would need to maintain, restore, or replace the cultural practices that produced them.

The river is but a long chain of flooded meadows.

APRIL 7, 1853

It is a fresh, cool summer morning . . . the Great Meadows have a slight bluish misty tinge in part; elsewhere a sort of hoary sheen like a fine downiness, inconceivably fine and silvery far away,—the light reflected from the grass blades, a sea of grass hoary with light, the counterpart of the frost in spring. As yet no mower has profaned it; scarcely a footstep since the waters left it. Miles of waving grass adorning the surface of the earth.

JULY 2, 1851

As I wade through the middle of the meadows in sedge up to my middle and look afar over the waving and rustling bent tops of the sedge (all are bent northeast by the southwest wind) toward the distant mainland, I feel a little as if caught by a rising tide on flats far from the shore. I am, as it were, cast away in the midst of the sea. It is a level sea of waving and rustling sedge about me. The grassy sea. You feel somewhat as you would if you were standing in water at an equal distance from the shore.

AUGUST 3, 1859

I paddle directly across the meadow, the river is so high, and land east of the elm on the third or fourth row of potatoes. The water makes more show on the meadows than yesterday, though hardly so high, because the grass is more flatted down. I easily make my way amid the thin spires. Almost every stem which rises above the surface has a grasshopper or caterpillar upon it. Some have seven or eight grasshoppers, clinging to their masts, one close and directly above another, like shipwrecked sailors, now the third or fourth day exposed. Whither shall they jump? It is a quarter of a mile to shore, and countless sharks lie in wait for them. They are so thick that they are like a crop which the grass bears; some stems are bent down by their weight. This flood affects other inhabitants of these fields than men; not only the owners of the grass, but its inhabitants much more. It drives them to their upper stories,—to take refuge in the rigging. Many that have taken an imprudent leap are seen struggling in the water. How much life is drowned out that inhabits about the roots of the meadow-grass! How many a family, perchance, of short-tailed meadow mice has had to scamper or swim!

AUGUST 25, 1856

Meadow haying has commenced. There is no pause between the English [upland] and meadow haying. There are thousands of yellow butterflies on the pontederia [pickerel weed] flowers, and of various colors on the buttonbush. In the Sudbury meadows are dense fields of pipes [horsetails] three feet high bordering the river. The common large rush, flowering at top, makes black-looking squads there. The fields of pontederia are in some places four or five rods wide and almost endless, but, crossing from side to side on shore, are the open white umbels of the hemlock [water hemlock, *Cicuta maculata*] and now thesium begins to show.

JULY 18, 1853

The Great Meadows present a very busy scene now. There are at least thirty men in sight getting the hay, revealed by their white shirts in the distance, the farthest mere specks, and here and there great loads of hay, almost concealing the two dor-bugs that draw them—and horse racks [*sic*] pacing regularly back and forth. It is refreshing to behold and scent even this wreck of the meadow-plants. Here is a man sedulously cocking up great heaps composed almost alone of flowering fern, yet perfectly green. Here are many owners side by side, each taking his slice of the great meadow. The mower fixes bits of newspaper to stakes in straight lines across the meadow to guide him, lest he cut over his bounds. The completion of haying might be celebrated by a farmers' festival.

JULY 29, 1853

I find that we are now in the midst of the meadow-haying season, and almost every meadow or section of a meadow has its band of half a dozen mowers and rakers, either bending to their manly work with regular and graceful motion or resting in the shade, while the boys are turning the grass to the sun. I passed as many as sixty or a hundred men thus at work to-day. They stick up a twig with the leaves on, on the river's brink, as a guide for the mowers, that they may not exceed the owner's bounds. I hear their scythes cronching the coarse weeds by the river's brink as I row near. The horse or oxen stand near at hand in the shade on the firm land, waiting to draw home a load anon. I see a platoon of three or four mowers, one behind the other, diagonally advancing with regular sweeps across the broad meadow and ever and anon standing to whet their scythes. Or else, having made several bouts, they are resting in the shade on the edge of the firm land. In one place I see one sturdy mower stretched on the ground amid his oxen in the shade of an oak trying to sleep; or I see one wending far inland with a jug to some well-known spring.

AUGUST 5, 1854

A great part of the farmers of Concord are now in the meadows, and toward night great loads of hay are seen rolling slowly along the river's bank,—on the firmer ground there,—and perhaps fording the stream itself, toward the distant barn, followed by a troop of tired haymakers.

AUGUST 7, 1854

Now Lee and his men are returning to their meadow-haying after dinner, and stop at the well under the black oak in the field. I too repair to the well when they are gone, and taste the flavor of black strap [a licorice-tasting and slightly alcoholic beverage] on the bucket's edge. As I return down-stream, I see the haymakers now raking with hand or horse rakes into long rows or loading, one on the load placing it and treading it down, while others fork it up to him; and others are gleaning with rakes after the forkers. All farmers are anxious to get their meadow-hay as soon as possible for fear the river will rise.

AUGUST 5, 1854

Found a good stone jug, small size, floating stopple up. I drew the stopple and smelled, as I expected, molasses and water, or something stronger (black-strap?), which it *had* contained. Probably some meadow-haymakers' jug left in the grass, which the recent rise of the river has floated off. It will do to put with the white pitcher I found and keep flowers in. Thus I get my furniture.

NOVEMBER 9, 1855

Think what the farmer gets with his hay,—what his river-meadow hay consists of,—how much of fern and osier [dogwood shrub] and sweet-gale and *Polygonum hydropiperoides* and rhexia (I trust the cattle love the scent of it as well as I) and lysimachia, etc., etc., and rue, and sium and cicuta. In a meadow now being mown I see that the ferns and small osiers are as thick as the grass. If modern farmers do not collect elm and other leaves for their cattle, they do thus mow and cure the willows, etc., etc., to a considerable extent, so that they come to large bushes or trees only on the edge of the meadow.

AUGUST 3, 1856

David Heard says that the cattle liked the pipes so well that they distinguished their rustle from that of other grass as he was bringing them to them, and were eager to get them.

AUGUST 14, 1859

Mr. Hosmer is loading hay in his barn. It is meadow-hay, and I am interested in it chiefly as a botanist. If meadow-hay is of less worth in the market, it is more interesting to the poet. In this there is a large proportion of *Osmunda regalis* [royal fern].

JANUARY 5, 1858

Unfortunate those who have not got their hay. I see them wading in overflowed meadows and pitching the black and mouldy swaths about in vain that they may dry.

AUGUST 1, 1856

We have passed men at work in the water a foot or more deep, saving the grass they had cut, and now we enter the broader Sudbury meadows . . . Many tons stand cocked up, blackened and lost, in the water, and probably (?) they will not get the grass now standing.

AUGUST 19, 1853

The river meadows are now so dry that E. Wood is burning the Mantatuket one.

APRIL 27, 1860

Saw a meadow said to be still on fire after three weeks; fire had burned holes one and a half feet deep; was burning along slowly at a considerable depth.

AUGUST 31, 1854

I see how meadows were primitively kept in the state of meadow by the aid of water,—and even fire and wind . . . Maples, alders, birches, pitch and white pines are slow to spread into it. I have named them in the order of their slowness. The last are the foremost,—furthest into the meadow,—but they are sickly-looking. You may say that it takes a geological change to make a wood-lot there.

OCTOBER 22, 1860

What is the use, in Nature's economy, of these occasional floods in August? Is it not partly to preserve the meadows open?

AUGUST 25, 1856

I began to survey the meadow there early, before Miles's new mill had been running long this Monday morning and flooded it, but a great stream of water was already rushing down the brook, and it almost rose over our boots in the meadow before we had done.

APRIL 28, 1856

It is pleasant to paddle over the meadows now, at this time of flood, and look down on the various meadow plants, for you can see more distinctly quite to the bottom than ever. A few sedges are very common and prominent, one, the tallest and earliest, now gone and going to seed, which I do not make out, also the *Carex scoparia* and the *C. stellulata.* How will the water affect these plants, standing thus long over them? The head of every sedge that now rises above the surface is swarming with insects which have taken refuge from the flood on it,—beetles, grasshoppers, spiders, caterpillars, etc. How many must have been destroyed! No doubt thousands of birds' nests have been destroyed by the flood,—blackbirds', bobolinks', song sparrows', etc. I see a robin's nest high above the water with the young just dead and the old bird in the water, apparently killed by the abundance of rain, and afterward I see a fresh song sparrow's nest which has been flooded and destroyed. Two sternothaerus which I smell of have no scent to-day.

JUNE 16, 1858

[Minott] Said that Clark (Daniel or Brooks) asked him the other day what made so many young alders and birches and willows spring up in the river meadows of late years; it didn't use to be so forty or fifty years ago; and he told him that in old times, when they were accustomed to take something strong to drink, they didn't stand for such shrubs but mowed all clear as they went, but now, not feeling so much energy for want of the stimulant, when they came to a bush, though no bigger than a pipe-stem, they mowed all round it and left it standing.

SEPTEMBER 30, 1856

Looking from this side, the meadow appears to be filled almost exclusively with wool-grass, yet very little has any culm or has blossomed this year. I notice, however, one tract, in the midst of the rest, an oblong square with perfectly straight sides, reaching from the upland toward the river, where it has quite generally blossomed and the culms still stand as high as my head. This, plainly, is because the land of a particular proprietor has been subjected to a peculiar treatment.

AUGUST 26, 1858

The mower, perchance, cuts some plants which I have never seen in flower.

JULY 12, 1852

The mower on the river meadows, when [he] comes to open his hay these days, encounters some overgrown water adder [snake] full of young (?) and bold in defense of its progeny, and tells a tale when he comes home at night which causes a shudder to run through the village,—how it came at him, and he ran, and it pursued and over-took him, and he transfixed it with a pitchfork and laid it on a cock of hay, but it revived and came at him again. This is the story he tells in the shops at evening. The big snake is a sort of fabulous animal. It is always as big as a man's arm and of indefinite length. Nobody knows exactly how deadly its bite, but nobody is known to have been bitten and recovered. Irishmen introduced into these meadows for the first time, on seeing a snake, a creature which they have seen only in pictures before, lay down their scythes and run as if it were the evil one himself, and cannot be induced to return to their work. They sigh for Ireland, where they say there is no venomous thing that can hurt you.

AUGUST 5, 1853

As we went down the bank through A. Hosmer's land we saw great cakes, and even fields of ice, lying up high and dry where you would not suspect otherwise that water had been. Some have much of the withered pickerel-weed, stem and leaves, in it, causing it to melt and break up soon in the sun. I saw one cake of ice, six inches thick and more than six feet in diameter, with a cake of meadow of exactly equal dimensions attached to its under side, exactly and evenly balanced on the top of a wall in a pasture forty rods from the river, and where you would not have thought the water ever came. We saw three white maples about nine inches in diameter which had been torn up, roots and sod together, and in some cases carried a long distance. One quite sound, of equal size, had been bent flat and broken by the ice striking it some six or seven feet from the ground. Saw some very large pieces of meadow lifted up or carried off at mouth of G. M. Barrett's Bay. One measured seventy-four by twenty-seven feet. Topped with ice almost always, and the old ice still beneath. In some cases the black, peaty soil thus floated was more than one and a half feet thick, and some of this last was carried a quarter of a mile without trace of ice to buoy it, but probably it was first lifted by ice. Saw one piece more than a rod long and two feet thick of black, peaty soil brought from I know not where. The edge of these meadow-crusts is singularly abrupt, as if cut with a turf-knife.

MARCH 1, 1855

I think the meadow is lifted in this wise: First, you have a considerable freshet in midwinter, succeeded by severe cold before the water has run off much. Then, as the water goes down, the ice for a certain width on each side the river meadows rests on the ground, which freezes to it. Then comes another freshet, which rises a little higher than the former. This gently lifts up the river ice, and that meadow ice on each side of it which still has water under it, without breaking them, but overflows the ice which is frozen to the bottom. Then, after some days of thaw and wind, the latter ice is broken up and rises in cakes, larger or smaller with or without the meadow-crust beneath it, and is floated off before the wind and current till it grounds somewhere, or melts and so sinks, frequently three cakes one upon another, on some swell in the meadow or the edge of the upland. The ice is thus with us a wonderful agent in changing the aspect of the surface of the river-valley. I think that there has been more meadow than usual moved this year, because we had so great a freshet in midwinter succeeded by severe cold, and that by another still greater freshet before the cold weather was past.

MARCH 1, 1855

A. Wright says that about forty years ago an acre of meadow was carried off at one time by the ice on the Colburn place. D. Clark tells me he saw a piece of meadow, on his part of the Great Meadows, five or six rods square, which had been taken up in one piece and set down again a little distance off. I observe that where there is plowed ground much of it has been washed over the neighboring grass ground to a great distance, discoloring it.

MARCH 2, 1855

There are so many sportsmen out that the ducks have no rest on the Great Meadows, which are not half covered with water. They sit uneasy on the water, looking about, without feeding, and I see one man endeavor to approach a flock crouchingly through the meadow for half a mile, with india-rubber boots on, where the water is often a foot deep. This has been going on, on these meadows, ever since the town was settled, and will go on as long as ducks settle here.

MARCH 25, 1858

Stone Walls and Other Fences

> Excepting those fences which are mere boundaries of individual property, the walker can generally perceive the reason for those which he is obliged to get over. This [stone] wall runs along just on the edge of the hill and following all its windings, to separate the more level and cultivatable summit from the slope, which is only fit for pasture or wood-lot, and that other wall below divides the pasture or wood-lot from the richer low grass ground or potato-field, etc. Even these crooked walls are not always unaccountable and lawless.
>
> JULY 12, 1852

Few sights capture the extent of the transformation that has occurred in landscape character and human activity in New England as well as that of an ancient stone wall snaking across a forested hillside. Composed of rough stones plucked from the ground and piled in straight to ambling lines, these long-abandoned fences often have a lichen, moss, and leaf cover that make them appear to be a natural part of the landscape. Yet with some imagination on the viewer's part, these walls tell stories of an earlier civilization's efforts to clear the land, to tame a forested wilderness, and to improve plowed fields and open pastures. With some additional imagining we can erase the forest and trees from our view and recover the farmlands that extended across valleys and hills and separated the occasional woodlot from a house or road, before the neglect of the fields allowed trees to invade, grow up, and engulf the aging walls.

Although stone walls are a common sight as we walk or drive through New England, most of us know little about them. We may regard them simply as old fences, giving little thought to their extent and distribution in the land, the wide variation in their type and pattern, and the role they may have played in the landscape scheme of their day. Why and when were they constructed? What purposes did they actually serve, and can we detect this purpose in their form and pattern? Were they the common mode of fencing, or, by virtue of their massive and durable stoniness, are they only the survivors of a much more extensive system of fences? How did farmers decide on the placement of these walls and other fencing? What is the explanation of their sometimes unusual and complicated patterns—rectangular, acute and angular, or aligned in parallel rows, some occasionally terminating abruptly?

These derelict systems of walls and fences are common across most parts of the New England landscape, and in many places they are extensive. For example, in Petersham, Massachusetts, a township six by six miles in size and one of the few localities where the pattern of old stone walls has been comprehensively mapped, we know that they form an intricate but variable pattern that extends over 280 miles in total length. In the area of Concord where Thoreau roamed, stone walls are somewhat less common because the soil is less rocky and large areas are covered by sandy glacial outwash. However, in the Easterbrook country of north Concord and Carlisle, which Thoreau identified as one of three wild and rough areas of his haunts, we find a dense and complex system of old walls and narrow trails winding through the woods, where rugged pastures were once common.

Broadly speaking, there are two predominant types of stone walls. "Single" walls consist of simple lines of typically large stones that usually enclosed pasture land. Since these fences served primarily to contain livestock and to separate lands of different use and ownership, farmers seldom bothered to construct them in elaborate form or in size beyond that demanded by their limited functional need. A second, much broader "double" stone wall often consists of well-formed parallel lines of larger stones with the intervening space filled with many small stones. These massive walls surround former cultivated fields where the collection of stones following plowing was an annual activity for all farmers, who added to and continually shaped these walls over long periods of time and through repeated years of use.

The area of the original fields bounded by either type of wall may be a fraction of an acre or more than 5 acres, but generally they are small, matching the fine scale of the dissected and varied topography of the New England uplands and indicating the ability of farmers working with hand tools and animals to fit their activity closely to the land. On the most marginal lands where farming persisted for only a brief period, the sight of piles or rudimentary heaps of stones suggests that the inhabitants didn't occupy the area long enough to bother to construct fences. Vast collections of small rocks were also spread across extensive low and wet areas or piled into the corners of fields. Here they document the commonsense ability of farmers to put the apparently useless by-product of field improvement to good use in improving drainage on wet soils and providing easy access over seasonally inundated areas. Although invisible to the modern traveler, many a dirt road crossing a stream or woodland swale is actually underlain by the rocks removed from an adjacent upland area during the decades when it served as productive cropland. The functional applications of stones and walls are equally diverse. Parallel lines of stones often enclose former laneways, or chutes, that were used to direct or confine the movement of cattle; walls bordering roads are frequently broken by gateways edged on either side by very large upright stones.

Despite our ability to interpret a few of the patterns and uses of stone walls retrospectively, for a true understanding of fences we need a contemporary perspective of how they fit and functioned in the living landscape. For Thoreau, stone walls were only one element in an expansive network of fencing that added structure to the countryside. From valley bottom to hillcrest, fences of various kinds broke the undulating landscape, "framing" and "confining" it into different types of functional parcels. Thoreau notes that fences, rather than being placed haphazardly across the countryside, were normally situated along natural breaks in the landscape that were defined by changes in the soil or topography. At the bottom of a slope a wall might separate the rich, low meadows and grasslands from the rough and stony hillslope pasture above. Toward the top, the pasture might be bounded by a second wall and thereby separated from the level summit, which had been cultivated for crops. Although it may be difficult in today's landscape to understand their placement, Thoreau confirms that each fence served a very definite purpose—the walls "may be crooked, but they are not lawless." In addition to controlling the movement of animals and separating areas of different use, fences and walls also served to identify ownerships and townships. A regular responsibility of all elected selectmen in nineteenth-century New England was to "perambulate" and confirm the town boundaries with their counterparts from the adjoining township, an event that Thoreau notes as occurring every three years in Massachusetts. At township corners the selectmen would place rocks on piles or walls and would occasionally mark the event with a stone chiseled and lettered to record the date and the initials of the officials.

Equally interesting are the diverse types of fencing that Thoreau encountered, most of which have not survived in the modern landscape. Stone walls, for example, are seldom described by Thoreau as simple, separate structures but more often as surmounted with wooden "riders"—poles and posts that ran above and parallel to the wall to increase its height and improve the control of livestock. Here the stones added many advantages: their uneven and rocky surface dissuaded cows and horses from leaning and pushing against the fence, as they are prone to do with simple rails or modern wire fences, and the wall provided a dry and well-drained substrate to hold the posts away from the damp soil, in which they would quickly rot away. Wooden rail fences, of which there were many types, had either wooden or stone posts. Occasionally these and other fences were paralleled

by an adjacent ditch, which served as a further deterrent to livestock; this was analogous to the ancient system of walls and ditches occurring across the British and European countryside. Perhaps most unusual to us are the widespread root fences and the turf and sod walls that Thoreau saw constructed by recent Irish immigrants. In particular, the fences made of tree roots tipped up on their sides and arranged in interlacing rows fascinated Thoreau because of their raw beauty and stark brightness, their intricate pattern of gnarled interconnected parts, and their durability over decades or perhaps a century. These and each of the diverse types of fencing took time and care to maintain; mending and repairing them formed a regular part of a farmer's chores. Subsequently, as the New England farmers turned away from the land and the forests came back, the walls settled through neglect, the wooden riders fell and decayed, the ditches filled and collapsed, and even the durable roots returned to the soil, leaving only the stone walls or the upright granite posts as an incomplete record of the past.

In his discussion of fields and fences, Thoreau sheds light on another important though inadvertent ecological function of these boundaries of wood, stone, and root: their role in harboring a diversity of plants and small mammals. In a New England landscape of fields closely cropped by grazing animals, tilled for grains and vegetables, or cut repeatedly for hay, the margins of fields and fencerows provided protection for small mammals and a haven for the plants that might otherwise occupy the land. The native plants of the fields, notes Thoreau, retreated back to the sides of walls and fences, where a rich flora persisted. Even the most complete mowing or plowing by a careful farmer generally missed a marginal strip adjacent to the field's edge. Along these edges there developed a thick hedge or coppice of trees, shrubs, and herbs, and these fencerows of vegetation and stone, in turn, harbored a diverse array of animals. They also served as the source of the plants that eventually spread out onto fields as they were used less intensively and then abandoned. The small size of the fields and this tendency for fencerows to maintain and concentrate a wide range of plants are among the reasons that native plants and forests recolonized the landscape so rapidly.

Thus, as we read Thoreau's descriptions of the types and arrangements of fencing and the role they played in land-use patterns, we come away with a greater ability to read and understand our modern landscape. An old wall in the woods very likely separates two pieces of land that were used differently in the past, either as a consequence of ownership patterns or because of original differences in soils and vegetation. The modern forests on either side of a wall often reflect these different histories through striking contrasts in species composition or forest structure. Frequently the walls are overtopped by a parallel row of large spreading trees that once grew as open fencerow trees, offering shade to lolling cattle. The surrounding forests are often composed of smaller, narrow-topped trees that invaded the abandoned fields, grew up densely packed, and continue to reach narrowly upward for light. More subtle differences may be noticed on close inspection. For example, nut-bearing trees—oaks, hickories, chestnuts—may predominate along the walls where their seeds were stored by chipmunks and mice seeking shelter between the stones and in the fencerow, whereas species with wind-dispersed seeds such as pine, birch, maples, and ash may have spread easily across the former old fields, where they continue to dominate. These differences are legacies of a history that modern observers can still read in the land, its stone walls, and its fences.

We are never prepared to believe that our ancestors lifted large stones or built thick walls. I find that I must have supposed that they built their bank walls of such as a single man could handle. For since we have put their lives behind us we can think of no sufficient motive for such exertion. How can their works be so visible and permanent and themselves so transient? When I see a stone which it must have taken many yoke of oxen to move, lying in a bank wall which was built two hundred years ago, I am curiously surprised, because it suggests an energy and force of which we have no memorials. Where are the traces of the corresponding moral and intellectual energy? I am not prepared to believe that a man lived here so long ago who could elevate into a wall and properly aline a rock of great size and fix it securely,—such an Archimedes.

1850 [DAY UNKNOWN]

I am amused to see from my window here how busily man has divided and staked off his domain. God must smile at his puny fences running hither and thither everywhere over the land.

FEBRUARY 20, 1842

In that little pasture of Potter's under the oak, I am struck with the advantage of the fence in landscapes. Here is but a half-acre inclosed, but the fence has the effect of confining the attention to this little undulation of the land and to make you consider it by itself, and the importance of the oak is proportionally increased. This formation of the surface would be lost in an unfenced prairie, but the fence, which nearly enough defines it, frames it and presents it as a picture.

APRIL 17, 1852

The farmers on all sides are mending their fences and turning out their cows to pasture. You see where the rails have been newly sharpened, and the leafing birches have been cut and laid over gaps in the walls as if old fences were putting forth leaves.

MAY 12, 1853

The bright-yellow sulphur lichens on the [stone] walls of the Walden road look novel, as if I had not seen them for a long time. Do they not require cold as much as moisture to enliven them? What surprising forms and color! Designed on every natural surface of rock or tree. Even stones of smaller size which make the walls are so finished, and piled up for what use? How naturally they adorn our works of art! See where the farmer has set up his post-and-rail fences along the road. The sulphur lichen has, as it were, at once leaped to occupy the northern side of each post, as in towns handbills are pasted on all bare surfaces, and the rails are more or less gilded with them as if it had rained gilt.

DECEMBER 6, 1859

The [stone] wall by the road at the bars north of Cyrus Smith's chestnut grove is very firmly bound together by the *Rhus Toxicodendron* [poison ivy] which has overrun it, for twenty feet in length. Would it not be worth the while to encourage its growth for this purpose, if you are not afraid of being poisoned?

MAY 9, 1860

More sensible is a rod of stone wall that bounds an honest man's field than a hundred-gated Thebes that has mistaken the true end of life, that places hammered marble before honesty.

JUNE 26, 1852

This appears to be a very ancient custom, and I find that this word "perambulation" has exactly the same meaning that it has at present in Johnson and Walker's dictionary. A hundred years ago they went round the towns of this State every three years. And the old selectmen tell me that, before the present split stones were set up in 1829, the bounds were marked by a heap of stones, and it was customary for each selectman to add a stone to the heap.

SEPTEMBER 12, 1851

It is only necessary that man should start a fence that Nature should carry it on and complete it. The farmer cannot plow quite up to the rails or wall which he himself has placed, and hence it often becomes a hedgerow and sometimes a coppice [a dense growth of sprouting trees or shrubs].

FEBRUARY 12, 1851

How important the hazelnut to the ground squirrel! They grow along the walls where the squirrels have their homes. They are the oaks that grow before their doors. They have not far to go to their harvesting. These bushes are generally stripped, but isolated ones in the middle of fields, away from the squirrel-walks, are still full of burs. The wall is highway and rampart to these little beasts. They are almost inaccessible in their holes beneath it, and on either side of it spring up, also defended by the wall, the hazel bushes on whose fruit the squirrels in a great measure depend.

SEPTEMBER 3, 1858

Most fields are so completely shorn now that the walls and fence-sides, where plants are protected, appear unusually rich. I know not what aspect the flowers would present if our fields and meadows were untouched for a year, if the mower were not permitted to swing his scythe there. No doubt some plants contended long in vain with these vandals, and at last withdrew from the contest. About these times some hundreds of men with freshly sharpened scythes make an irruption into my garden when in its rankest condition, and clip my herbs all as close as they can, and I am restricted to the rough hedges and worn-out fields which had little to attract them, to the most barren and worthless pastures.

JULY 29, 1853

When I ran out [surveyed] the boundary lines of this lot, I could commonly distinguish the line, not merely by the different growth of wood, but often by a kind of ditch which I think may have been produced by the plow, which heaped up the soil along the side of the field when it was cultivated. I could also detect trees variously bent and twisted, which probably had made part of a hedge fence when young, and others which were scarred by the fencing-stuff that had been fastened to them.

NOVEMBER 26, 1859

When I was surveying Shattuck's Merrick's [*sic*] pasture fields the other day, McManus, who was helping me, said that they would be worth a hundred or two hundred dollars more if it were not for the willow-rows which bound and separate them, for you could not plow parallel with them within five rods on account of the roots, you must plow at right angles with them. Yet it is not many years since they were set out, as I remember. However, there should be a great amount of root to account for their wonderful vivaciousness, making seven or eight feet in a year when trimmed.

JANUARY 16, 1857

I see that E. Wood has sent a couple of Irishmen, with axe and bush-whack, to cut off the natural hedges of sumach, Roxbury wax-work, grapes, etc., which have sprung up by the walls on this hill farm, in order that his cows may get a little more green. And they have cut down two or three of the very rare celtis [hackberry, *Celtis occidentalis*] trees, not found anywhere else in town. The Lord deliver us from these vandalic proprietors! The botanist and lover of nature has, perchance, discovered some rare tree which has sprung up by a farmer's wall-side to adorn and bless it, sole representative of its kind in these parts. Strangers send for a seed or a sprig from a distance, but, walking there again, he finds that the farmer has sent a raw Irishman, a hireling just arrived on these shores, who was never there before,—and, we trust, will never be let loose there again,—who knows not whether he is hacking at the upas tree or the Tree of Knowledge, with axe and stub-scythe to exterminate it, and he will know it no more forever.

SEPTEMBER 28, 1857

I saw an Irishman building a bank of sod where his employer had contemplated building a bank wall, piling up very neatly and solidly with his spade and a line the sods taken from the rear, and coping the face at a very small angle from the perpendicular, intermingling the sods with bushes as they came to hand, which would grow and strengthen the whole. It was much more agreeable to the eye, as well as less expensive, than stone would have been, and he thought that it would be equally effective as a fence and no less durable. But it is true only experience will show when the same practice may be followed in this climate and in Ireland,—whether our atmosphere is not too dry to admit of it. At any rate it was wise in the farmer thus to avail himself of any peculiar experience which his hired laborer possessed.

JUNE 29, 1851

I am attracted by a fence made of white pine roots. There is, or rather was, one (for it has been tipped into the gutter this year) on the road to Hubbard's Bridge which I can remember for more than twenty years. It is almost as indestructible as a [stone] wall and certainly requires fewer repairs. It is light, white, and dry withal, and its fantastic forms are agreeable to my eye. One would not have believed that any trees had such snarled and gnarled roots. In some instances you have a coarse network of roots as they interlaced on the surface perhaps of a swamp, which, set on its edge, really looks like a fence, with its paling crossing at various angles, and root repeatedly growing into root,—a rare phenomenon above ground,—so as to leave open spaces, square and diamond-shaped and triangular, quite like a length of fence. It is remarkable how white and clean these roots are, and that no lichens, or very few, grow on them; so free from decay are they. The different branches of the roots continually grow into one another, so as to make grotesque figures, sometimes rude harps whose resonant strings of roots give a sort of musical sound when struck, such as the earth spirit might play on.

NOVEMBER 11, 1850

John Potter told me that those root fences on the Corner road were at least sixty or seventy years old.

AUGUST 17, 1851

I admire those old root fences which have almost entirely disappeared from tidy fields,—white pine roots got out when the neighboring meadow was a swamp,—the monuments of many a revolution. These roots have not penetrated into the ground, but spread over the surface, and, having been cut off four or five feet from the stump, were hauled off and set up on their edges for a fence. The roots are not merely interwoven, but grown together into solid frames, full of loopholes like Gothic windows of various sizes and all shapes, triangular and oval and harp-like, and the slenderer parts are dry and resonant like harp-strings. They are rough and unapproachable, with a hundred snags and horns which bewilder and balk the calculation of the walker who would surmount them. The part of the trees above ground presents no such fantastic forms. Here is one seven paces, or more than a rod, long, six feet high in the middle, and yet only one foot thick, and two men could turn it up, and in this case the roots were six or nine inches thick at the extremities. The roots of pines growing in swamps grow thus in the form of solid frames or rackets, and those of different trees are interwoven with all so that they stand on a very broad foot and stand or fall together to some extent before the blasts, as herds meet the assault of beasts of prey with serried front. You have thus only to dig into the swamp a little way to find your fence,—post, rails, and slats already solidly grown together and of material more durable than any timber. How pleasing a thought that a field should be fenced with the roots of the trees got out in clearing the land a century before! I regret them as mementoes of the primitive forest. The tops of the same trees made into fencing-stuff would have decayed generations ago.

DECEMBER 23, 1855

Among old stumps I have not named those white pine ones used as fences with their roots. I think that some of these must be older than any left in the ground. I remember some on the Corner road, which apparently have not changed for more than thirty years, and are said to be ninety years old. Lying thus high and dry, they are almost indestructible, and I can still easily count the rings of many of these. I count one hundred and twenty-six rings on one this afternoon, and who knows but it is a hundred years since it was cut? They decay much faster left upright in the ground than lying on their sides on the surface, supposing it open land in both cases.

Perhaps these great pine roots which grew in a swamp were provided with some peculiar quality by which to resist the influence of moisture and so endure the changes of the weather.

OCTOBER 31, 1860

There sits a stone-mason, splitting Westford granite for fence-posts. Egypt has perchance taught New England something in this matter. His hammer, his chisels, his wedges, his shims or half-rounds, his iron spoon,—I suspect that these tools are hoary with age as with granite dust. He learns as easily where the best granite comes from as he learns how to erect that screen to keep off the sun. He knows that he can drill faster into a large stone than a small one, because there is less jar and yielding. He deals in stone as the carpenter in lumber. In many of his operations only the materials are different. His work is slow and expensive. Nature is here hard to be overcome. He wears up one or two drills in splitting a single stone. He must sharpen his tools oftener than the carpenter. He fights with granite. He knows the temper of the rocks. He grows stony himself. His tread is ponderous and steady like the fall of a rock. And yet by patience and art he splits a stone as surely as the carpenter or woodcutter a log. So much time and perseverance will accomplish. One would say that mankind had much less moral than physical energy, that any day you see men following the trade of splitting rocks, who yet shrink from undertaking apparently less arduous moral labors, the solving of moral problems. See how surely he proceeds. He does not hesitate to drill a dozen holes, each one the labor of a day or two for a savage; he carefully takes out the dust with his iron spoon; he inserts his wedges, one in each hole, and protects the sides of the holes and gives resistance to his wedges by thin pieces of half-round iron (or shims); he marks the red line which he has drawn, with his chisel, carefully cutting it straight; and then how carefully he drives each wedge in succession, fearful lest he should not have a good split!

SEPTEMBER 11, 1851

Most plowed fields are quite bare, but I am surprised to find behind the walls on the south side, like a skulking company of rangers in ambuscade or regular troops that have retreated to another parallel, a solid column of snow six or eight feet deep. The wind, eddying through and over the wall, is scooping it out in fantastic forms,— shells and troughs and glyphs of all kinds.

DECEMBER 27, 1853

A Natural History of Woodlands

Woodlands and Sproutlands

It is an interesting inquiry what determines which species of these [trees] shall grow on a given tract. It is evident that the soil determines this to some extent. . . .

Secondly, ownership, and a corresponding difference of treatment of the land as to time of cutting, etc.,. . . .

Third, age, as, if the trees are one hundred years old, they may be chestnut, but if sprout-land are less likely to be; etc., etc., etc.

OCTOBER 17, 1860

In the nineteenth-century New England countryside, where agriculture was the predominant land use and woodlots were small and provided a diverse range of products, forests and human activity were inextricably linked. To most people, and particularly to Henry Thoreau and his rural acquaintances who roamed the landscape widely, each woodlot in the byways that they traveled was known intimately: its name, appearance, characteristic plant species, and its history of ownership and use. In fact, Thoreau claimed that he knew the character of every woodlot so well that he could identify the location of each owl that hooted in the night and the source of every large log that was sawn at the local mills or that passed by on a wagon or sled. In the vernacular, which appears in his journals and which he used with his acquaintances in Concord, the names of woodlots were based on a mixture of age-old place names, reference to past or current owners, proximity to notable landmarks, and his own enigmatic coinage.

Thoreau's forest terminology, on the other hand, is precise and, as is appropriate to the history and character of the nineteenth-century landscape of New England, strongly linked to that of the British countryside, where the woodlots were even less numerous than in New England and had been shaped over millennia within an agricultural landscape. Thoreau's evening readings, in addition to literary works by Wordsworth, De Quincey, Coleridge, and others, included the works of the English landscape architect William Gilpin, and he naturally compared the forests, woodland practices, and tree forms of the Concord landscape directly with those of England and Europe. Consequently, in developing a terminology suitable for his own woods, Thoreau borrowed from a classification of forest types employed by Gilpin and still in common use across Britain today. This system is based on the history and origin of woodlots, in which they are arranged in order of the intensity of human impact and their decreasing naturalness.

First in the hierarchy is the *primitive wood,* which has never been cut. Although Thoreau recognized that this type of natural forest was completely absent from Concord and most of southern New England, he believed that such stands still persisted in the nearby north-central Massachusetts town of Winchendon. He also asserted that primitive forests, though uncommon, were extremely valuable for the insights they could provide into natural processes and for the plant and animal species they harbored that were uncommon in less natural woods—two notions that are fiercely defended by enthusiasts of old-growth forests today. The desire to see landscapes composed of such primitive forest was certainly a major motivation for Thoreau's travels to the north woods of Maine.

Next in Thoreau's hierarchy of naturalness are the sites that have supported forest continuously through time, though occasionally cut, and have never been used for agriculture or burned intentionally; he called these sites *primitive woodland.* This category encompassed most of the cut-over forests and woodlots that existed in Thoreau's day at the height of New England agricultural activity and occupied the rocky, steep, and extremely sandy or wet sites that did not make good pasture or arable land. Such primitive woodland provided most of the fuel and timber in nineteenth-century New England. Thoreau further separated this category of woodland into *true second growth* that had been cut only once, and *woodlots,* such as those surrounding Walden Pond, that had been clear-cut repeatedly. He noted that the largest trees remaining in Concord during his day were all of second-growth origin, having regrown following a first cutting earlier in the town's history. *Interrupted* or *tamed woodlands* are the two terms Thoreau applied to forests that had established themselves on sites that had been cleared briefly for agriculture. The first generation of trees to develop on sites cleared for many years he called *new woods*. Thus, to use Thoreau's terms, the forest landscape that we have inherited from him consists predominantly of interrupted woodlands that have developed on old pastures and fields that became reforested as agriculture declined, and to a lesser extent of primitive woodland that was cut, but remained continuously forested. The true primitive wood or virgin forest that has come through 400 years of history untouched by human hands is very rare indeed.

The historical distinctions within Thoreau's system of forest classification are extremely useful to us in our interpretation of modern forests, and they were of critical importance to him because he recognized that the type and intensity of human use are major determinants of forest composition and structure, affecting the types of herbs and grasses in the ground cover as well as the age, species, and growth-form of trees in the overstory. More specifically, in his discussions of the factors that control the distribution of plant species and forests in his landscape, Thoreau singles out three: (1) soils and natural environment, (2) land use and ownership patterns, and (3) age or time since the last major disturbance. Forests and plant species, he notes, tend to be orderly. Each species has its own tendency to favor specific sites and environments and therefore "knows its ground." For example, among the common tree species in New England, Thoreau observed that the red, black, and white oaks predominated across the uplands; swamp white oak and red maple were more abundant on the low, moist swales; pitch pine was concentrated on droughty, sandy soils, ridgetops, and abandoned old fields; and white pine was fairly wide-ranging, occurring frequently with upland hardwoods as well as on wet soils. Thus, when given sufficient time to disperse through the landscape and grow to maturity, each species will gradually occupy and become dominant on the most suitable site. However, Thoreau also clearly recognized that the human tendency to disturb forests through cutting, burning, or clearing and to cultivate species where they would not occur naturally leads to disruption of this natural pattern. Consequently, he felt that each forest "is generally in a transition state to a settled and normal condition," and each woodlot has a history of cross-purposes—of nature constantly endeavoring to grow and sort out species across the landscape in accordance, for example, to the varying soil moisture of hilltops, valley bottoms, and swampland, and of man's interference and the impact of the occasional windstorm, fire, or pathogen.

In a heavily humanized landscape like the one that Thoreau lived in and that exists in most parts of the world today, a classification of forests and sites based on the history and intensity of human impact, the continuity of forest cover, and the age of the forest conveys a great deal of information. From a historical perspective, the most important criterion for understanding forests in Thoreau's day was whether the particular site had been continuously forested or whether the native flora had been removed at some point in the past through cutting followed by repeated grazing, burning, or plowing. Thoreau felt that a history of uninterrupted forest cover produced a greater naturalness in the current vegetation, a sentiment that we may agree with intuitively. However, he supported this notion with his repeated observations that some species, once removed from a site by agricultural practices, may have a difficult and very slow time re-invading and spreading back onto the site. In New England today we can see one of the best examples of this behavior in the distribution and characteristics of one our of most distinctive trees—the eastern hemlock *(Tsuga canadensis).* Although hemlock is a very long-lived and shade-tolerant species that grows both in the understory of forests and as a dominant tree in old-growth stands, it is strongly limited in its rate of dispersal, establishment, and spread into new forest areas and across the landscape. Consequently, once intense fire or agricultural clearance has removed hemlock from a site, it is very slow to reestablish. Hemlock therefore is often a very good indicator of forest history and continuity. Where large hemlocks grow abundantly, there is a strong likelihood that the site is one of Thoreau's "primitive woodlands" that have never been in agricultural use. Conversely, we would not expect to find many two-foot-diameter hemlocks in the areas indicated as open fields on the nineteenth-century maps of New England.

Thoreau also recognized, as we will see in his discussions of the abandonment of agricultural fields and forest succession, that there were many plant species that occurred preferentially in the "new woods" and "interrupted woodlands" that developed in the old fields as farmers neglected them. These species are either those that spread easily and establish rapidly on open sites (in contrast to hemlock), or they are weedy plants and grasses that are common in pastures and fields. Similarly, Thoreau observed that, once established, a surprising number of plants are able to persist in a forest as a consequence of their system of buried roots and rhizomes that provide an important source of nutrients and water for growth. Although these below-ground structures are highly resilient to repeated damage, they are effectively removed by the forest clearance and plowing associated with agriculture. Once the roots and rhizomes are eliminated from a site, the only means for these plants to reestablish is by seed. Thus Thoreau combined his observations on roots and sprouts with those on seeds and seedlings into an understanding of plant biology which he used, together with his reading of land-use history, to explain the landscape that surrounded him.

Although the general framework and approach that Thoreau presented for understanding forest history continue to be valid and very useful, many of the specific forest scenes that he described have disappeared through time and are unknown to us today. However, since forests are composed of long-lived trees and their progeny, they may retain a legacy of their past for decades or centuries after their previous uses have ceased. Thoreau's descriptions of nineteenth-century woodland practices thus provide us with insights on past conditions and uses that may be critical in interpreting the woodland characteristics that we experience on our own walks and travels through the modern countryside. For example, sproutlands ("coppices") were a ubiquitous sight in Thoreau's landscape, and they occur prominently in his descriptions of Concord's countryside. These young, dense woodlots had a history of repeated cutting and frequent burning, and they contained hardwood species that sprout prolifically, such as oaks, chestnut, birch, red maple, hickory, and cherry. Chestnut, which was extensively used for fence rails and poles, and alder, which was turned into charcoal for use in the manufacture of gunpowder, were two species that were cut on a very frequent basis. As a result they often formed dense coppices in which four to ten small sprouts grew in a compact, rounded clump that spread upward and outward from a basal "stool," or point of attachment with the root system. The sprout clumps used this extensive root system to grow very rapidly and formed impenetrable thickets of even-aged and even-sized stems for the first 15 to 30 years, or until cut again.

The need for small fencing or pole-size material that led to the ancient practice of coppicing is gone today, and we seldom see such thickets of dense sprouts. However, the legacy of this intensive cutting remains in the extensive "sprout hardwood forests" that continue to dominate most of the eastern United States landscape. These forests are composed of species of hardwood trees (oak, birch, maple, hickory, chestnut) that share a common ability to produce new sprouts when the main stem is cut or injured. Thus the common sight of multiple-stemmed trees across New

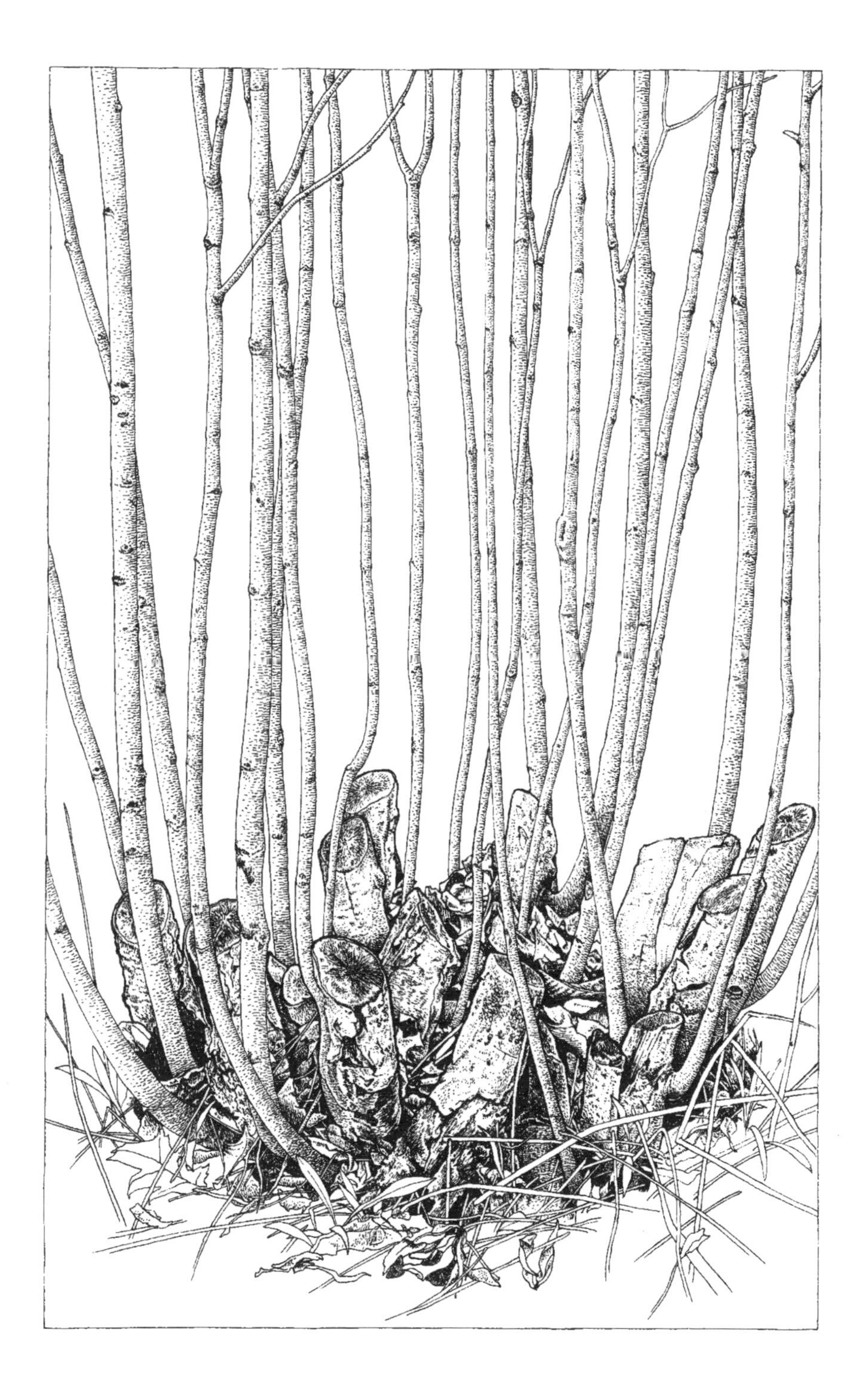

England is a direct consequence of sprouting behavior that was triggered by forest land-use practices 25 to 150 years earlier.

Even more distinctive and unusual than Thoreau's descriptions of tree coppices are his journal entries depicting the fencerow hedges and thickets that were created by the repeated cutting of shrubs such as shadbush and willow and the "pollarding" of elms and maples that was occasionally practiced along roadways. Pollarding is a traditional European practice of cutting the branches off a tree every three to five years at a height of 6 to 15 feet above the ground. Its purpose was identical to coppicing—to yield a regular supply of small-diameter sticks and poles as well as leaves, which served as fodder for domestic animals through the winter months. However, pollarding yielded one advantage over coppicing: the new shoots that were produced after cutting were formed atop trunks high above the ground and therefore out of the reach of grazing animals such as cows and horses. Thoreau clearly viewed the examples of pollarding that he saw as anomalies in New England, attributing them, as he did (occasionally erroneously) many unusual land-use practices, to the recently immigrated Irish.

The use of woodlots and forests as wooded pasture for cattle, sheep, and even hogs was a common scene in the nineteenth-century landscape, and Thoreau remarked on the distinctive forest appearance that this practice created as well as the characteristic sounds produced by large animals crashing their way through the woods. However, as woodland grazing continued through time and was accompanied by the cutting or thinning of trees, it inevitably led to the gradual elimination of small trees and saplings and the formation of a very open forest of rounded, widely spreading trees, grassy glades, trodden bare patches, and exposed rocks. In the extreme, woodland grazing eventually created open pastures with huge spreading oaks or chestnuts, under which the cattle huddled for shade in midsummer. This was a distinctive agrarian sight, and these large trees provided welcome cover for the solitary walker on a hot sunny day or during a sudden downpour. Thoreau, however, criticized the practice of woodland grazing because it led gradually but irrevocably to deterioration of the forest.

In our modern landscape we only occasionally encounter a scene of cows lolling and grazing among the trunks of a dense forest, or of a woodland path beaten down by the single-file movement of animals. This practice is generally restricted to smaller, less productive farms in the rural hill towns of New England because, by and large, farmers and foresters have come to agree with Thoreau that forest grazing is an unproductive and even harmful practice. However, the long-term consequences of this practice are still apparent in the openness of many forests and the remains of rocky paths created by hoofs in the woods. The extent to which such a history of intensive grazing affects the grasses, wildflowers, ferns, and trees that grow in a forest today is largely unknown because of our tendency to ignore, or our inability to know, the history of such past uses. However, we can observe a parallel scene in many modern forests as a result of a very different grazing animal. The deer that were largely absent from Thoreau's landscape have recently become so plentiful in many suburban and rural forest areas that they have created a novel version of this open, park-like forest. Like the nineteenth-century cows, deer will strip the understory clean of woody plants and many herbs and will leave a spacious and picturesque forest that is often quite low in plant diversity. In forests heavily grazed by deer the only common understory plants are the unpalatable ones, while many native herbs and tree seedlings are restricted to inaccessible sites such as rock ledges and cliffs where they are protected from the animal's reach. Overgrazing by deer has become a major problem for forest regeneration and for home landscapers and gardeners across the northeastern United States, including much of southern New England. This overabundance of deer, while providing a visual link with the cattle-grazed countryside of Thoreau, remains an unresolved and growing challenge for many New England communities.

Thoreau's descriptions are useful in many ways as we study the current landscape and think about our forests. His precise terminology and his approach to classifying forest types are relevant to the specific history of our land, and he discusses in some detail the relative importance of key factors that control forest structure and composition. The particular and unusual land-use practices that he describes are especially valuable as we seek to gauge the factors that have caused the major changes in New England forests and the relative abundance of particular plant species through time. Thoreau also reminds us of the beauty of a well-used forest in a strongly cultural landscape.

> I was struck by the orderly arrangement of the trees, as if each knew its own place . . . As if in the natural state of things, when sufficient time is given, trees will be found occupying the places most suitable to each, but when they are interfered with, some are prompted to grow where they do not belong and a certain degree of confusion is

produced. That is, our forest generally is in a transition state to a settled and normal condition.

NOVEMBER 5, 1860

We say of some plants that they grow in wet places and of others that they grow in desert places. The truth is that their seeds are scattered almost everywhere, but here only do they succeed.

OCTOBER 18, 1860

What Gilpin says about copses, glens, etc., suggests that the different places to which the walker resorts may be profitably classified. . . .

MARCH 31, 1852

1st. There is the primitive wood, woodland which was woodland when the township was settled, and which has not been cut at all. Of this I know of none in Concord. Where is the nearest? There is, perhaps, a large tract in Winchendon.
2d. Second growth, the woodland which has been cut but once,—true second growth. This country has been so recently settled that a large part of the older States is covered now with this second growth, and the same name is occasionally still applied, though falsely, to those wood-lots which have been cut twice or many more times. Of this second growth I think that we have considerable left, and I remember much more. These are our forests which contain the largest and oldest trees,—shingle pines (very few indeed left) and oak timber.
3d. Primitive woodland, *i.e.,* which has always been woodland, never cultivated or converted into pasture or grain-field, nor burned over intentionally. Of two kinds, first, that which has only been thinned from time to time, and secondly, that which has been cut clean many times over. A larger *copsewood.*
4th. Woodland which has been cleared one or more times, enough to raise a crop of grain on it, burned over and perhaps harrowed or even plowed, and suffered to grow up again in a year or two. Call this "interrupted woodland" or "tamed."
5th. *New woods,* or which have sprung up *de novo* on land which has been cultivated or cleared long enough to kill all the roots in it. (The 3d, 4th, and 5th are a kind of copsewood.)
6th. Artificial woods, or those which have been set out or raised from the seed, artificially.

OCTOBER 20, 1860

But the most interesting part of this walk [in the north-central Massachusetts town of Winchendon] was the three miles along the railroad between State Line and Winchendon Station. It was the best timbered region we saw, though its trees are rapidly falling [to the axe]. The railroad runs very straight for long distances here through a primitive forest. To my surprise I heard the *tea-lea* of the myrtle-bird here, as in Maine, and suppose that it breeds in this primitive wood. There was no house near the railroad but at one point, and then a quarter of a mile off. The red elder was in full bloom and filled the air with its fragrance. I saw some of the handsomest white pines here that I ever saw,—even in Maine,—close by the railroad. One by which I stood was at least three and a half feet in diameter at two feet from the ground, and, like several others about it, rose perfectly straight without any kind of limb to the height of sixty feet at least. What struck me most in these trees, as I was passing by, was not merely their great size, for they appeared less than they were, but their perfect perpendicularity, roundness, and apparent smoothness, tapering very little, like artificial columns of a new style. Their trunks were so very round that for that reason they appeared smoother than they were, marked with interrupted bands of light-colored lichens. Their regular beauty made such an impression that I was forced to turn aside and contemplate them. They were so round and perpendicular that my eyes slid off, and they made such an impression of finish and even polish as if they had an enamelled surface.

JUNE 4, 1858

Land where the wood has been cut off and is just beginning to come up again is called sprout land.

NOVEMBER 16, 1850

It is refreshing to walk over sprout-lands, where oak and chestnut sprouts are mounting swiftly up again into the sky, and already perchance their sere leaves begin to rustle in the breeze and reflect the light on the hillsides.

FEBRUARY 12, 1851

According to Gilpin, a copse is composed of forest trees mixed with brushwood, which last is periodically cut down in twelve or fourteen years.

MARCH 31, 1852

I forgot to say yesterday that several little groves of alders on which I had set my eye had been cut down the past winter. One in Trillium Woods was a favorite because it was so dense and regular, its outline rounded as if it were a moss bed; and another more than two miles from this, at Dugan's, which I went to see yesterday, was then being cut, like the former, to supply charcoal for powder. Dugan does most of this work about the town. The willow hedges by causeways are regularly trimmed and peeled. The small wood brings eight dollars a cord. Alders, also, and poplars are extensively used.

MARCH 25, 1853

If you examine a wood-lot after numerous fires and cuttings, you will be surprised to find how extremely vivacious are the roots of oaks, chestnuts, hickories, birches, cherries, etc. The little trees which look like seedlings of the year will be found commonly to spring from an older root or horizontal shoot or a stump. Those layers which you may have selected to transplant will be found to have too much of old stump and root underground to be removed. They have commonly met with accidents and seen a good deal of the world already. They have learned to endure and bide their time. When you see an oak fully grown and of fair proportions, you little suspect what difficulties it may have encountered in its early youth, what sores it has overgrown, how for years it was a feeble layer lurking under the leaves and scarcely daring to show its head above them, burnt and cut, and browsed by rabbits. Driven back to earth again twenty times,—as often as it aspires to the heavens. The soil of the forest is crowded with a mass of these old and tough fibres, annually sending up their shoots here and there. The underground part survives and holds its own, though the top meets with countless accidents; so that, although seeds were not to be supplied for many years, there would still spring up shoots enough to stock it. So with the old and feeble huckleberry roots. Nay, even the sedge *(Carex Pennsylvanica)* is already rooted in most woods, and at once begins to spread and prevail when the wood is cut, especially if a frost or fire keeps down the new wood.

OCTOBER 14, 1860

I detect the Irishman where the elms and maples on the causeway are cut off at the same height with the willows *to make pollards of!*

DECEMBER 23, 1855

The *new* woodlands, *i. e.,* forests that spring up where there were no trees before, are pine (or birch or maple), and accordingly you may see spaces of bare pasture sod between the trees for many years. But oaks, in masses, are not seen springing up thus with old sod between them. They form a sprout-land, or stand amid the stumps of a recent pine lot.

OCTOBER 19, 1860

Crossed on to the old Carlisle road by the house north of Isaiah Green's, and then across the road through the woods to the Paul Adams house by Bateman's Pond. Saw a hog-pasture of a dozen acres in the woods, with thirty or forty large hogs and a shelter for them at night, a half-mile east of the last house,—something rare in these days hereabouts.

JUNE 10, 1853

It is noon, and I hear the cattle crashing their way down the Cliff, seeking the shade of the woods. They climb like goats.

MAY 22, 1854

It is evident that the pasture oaks are commonly the survivors or relics of old oak woods . . . as an old oak wood is very gradually thinned out, it becomes open, grassy, and park-like, and very many owners are inclined to respect a few larger trees on account of old associations, until at length they begin to value them for shade for their cattle. These are oftenest white oaks. I think that they grow the largest and are the hardiest. This final arrangement is in obedience to the demand of the cow. She says, looking at the oak woods: "Your tender twigs are good, but grass is better. Give me a few at intervals for shade and shelter in storms, and let the grass grow far and wide between them."

NOVEMBER 5, 1860

You will see full-grown woods where the oaks and pines or birches are separated by right lines, growing in squares or other rectilinear figures, because different lots were cut at different times.

DECEMBER 7, 1856

Looking round, I observe at a distance an oak wood-lot some twenty years old, with a dense narrow edging of pitch pines about a rod and a half wide and twenty-five or thirty years old along its whole southern side, which is straight and thirty or forty rods long, and, next to it, an open field or pasture. It presents a very singular appearance, because the oak wood is broad and has no pines within it, while the narrow edging is perfectly straight and dense, and pure pine. It is the more remarkable at this season because the oak is all red and yellow and the pine all green. I understand it and read its history easily before I get to it. I find, as I expected, a fence separating the pines from the oaks, or that they belong to different owners.

OCTOBER 16, 1860

Saw some unusually broad chestnut planks, just sawed, at the mill. Barrett said that they came from Lincoln; whereupon I said that I guessed I knew where they came from, judging by their size alone, and it turned out that I was right. I had often gathered the nuts of those very trees and had observed within a year that they were cut down. So it appears that we have come to this, that if I see any peculiarly large chestnuts at the sawmill, I can guess where they came from, even know them in the log.

MAY 12, 1858

This wonderful activity of the squirrels in collecting and dispersing and planting nuts and acorns, etc., etc., every autumn is the more necessary since the trees on whose fruit they mainly live are not annual plants like the wheat which supplies *our* staff of life. If the wheat crop fails this year, we have only to sow more the next year, and reap a speedy harvest, but if the forests were to be planted only at intervals equal to the age of the trees, there would be danger, what with fires and blight and insects, of a sudden failure and famine. It is important that there be countless trees in every stage of growth,—that there be an annual planting, as of wheat. Consider the amount of work they have to do, the area to be planted!

NOVEMBER 2, 1860

It is evident that Nature's designs would not be accomplished if seeds, having been matured, were simply dropped and so planted directly beneath their parent stems, as many will always be in any case. The next consideration with her, then, after determining to create a seed, must have been how to get it transported, though to never so little distance,—the width of the plant, or less, will often be sufficient,—even as the eagle drives her young at last from the neighborhood of her eyrie,—for their own good, since there is not food enough there for all,—without depending on botanists, patent offices, and seedsmen. It is not enough to have matured a seed which will reproduce its kind under favorable conditions, but she must also secure it those favorable conditions. Nature has left nothing to the mercy of man. She has taken care that a sufficient number of every kind of seeds, from a cocoanut to those which are invisible, shall be transported and planted in a suitable place.

MARCH 22, 1861

Forest Land Use and Woodland Practices

This shows how heedlessly wood-lots are managed at present, and suggests that when one is cut (if not before) a provident husbandman will carefully examine the ground and ascertain what kind of wood is about to take the place of the old and how abundantly, in order that he may act understandingly and determine if it is best to clear the land or not. I have seen many a field perfectly barren for fifteen or twenty years, which, if properly managed, or only let alone, would naturally have yielded a crop of birch trees within that time.

SEPTEMBER 24, 1860

The history of woodland use emerges clearly from Thoreau's writings as a key factor in determining how forests changed from the time of New England settlement to his day, and from then to the present. How were these woodlands actually managed through time? How did these practices shape the nineteenth-century forests that Thoreau frequented, and how were they perceived by the naturalist of Concord? More vividly than any statistic, Thoreau's journal writings capture the reality of incessant woodland cutting that characterized the nineteenth-century New England landscape. In winter the sound of axes was everywhere, from earliest light until nightfall. Thoreau likened it to the chirping of birds, as though by its ubiquity chopping had become a part of nature. The cutting was so severe that he once described Concord as "sheared of its pride," thereby conveying both a stark image and the moral consequence of widespread and repeated clear-cutting. The impressions that emerge from the journals are corroborated by nineteenth-century census data, which document that most forests in Thoreau's day were composed of trees less than 30 years old and were cut every 10 to 40 years.

To Thoreau, it seemed that every forest of consequence was being felled and that every large tree he admired was eventually located, culled, and carted to the mill. By the 1850s the cutting had dramatically changed all of the woodland retreats that he had grown up with and knew best. The slopes above Walden Pond, originally thick with lofty pines and oaks and leading down to shores of vine-strewn coves, had been reduced to sproutland and the field openings in which Thoreau planted his beans. Large areas had been clear-felled, leaving pine, oak and chestnut stumps. Thoreau's written images of winter came to be dominated by woodchoppers, blazing fires of logging slash, teams of oxen waiting for their wooden loads, and long logs of white pine being sledded to the mills.

Unfortunately, the familiarity of Concord's citizens with their woodlots and their dependence on the woods for fuel, food, and many daily products did not mean that forests were generally managed in a thoughtful or careful manner. In fact, the range in approach and the widespread lack of foresight exhibited in their forest management were remarkable to Thoreau. For example, there was the unusual long-term attitude adopted by Minott, who, along with his sister, managed the family 10-acre woodlot deliberately and with careful attention so that it had faithfully yielded eight cords of firewood each year for more than 30 years. In his conversations with Thoreau, Minott expressed joy that he knew each tree on his land, its history, and its future use. In sharp contrast, there was the anonymous "owner" (ownership was a term that Thoreau often used contemptuously because he did not regard legal title as conveying any true rights to nature) who first clear-cut his woodlot and then proceeded to harrow and plant it with winter rye in order to eke out an additional cash crop of grain between the more lengthy forest rotations. Rather than working with nature and understanding his woods, this owner followed a short-sighted but direct path to failure. As he harvested logs from his woodlot, he concentrated on the wood alone and ignored the abundance of tree seedlings and sprouts that covered the ground and that represented the next generation of trees. By following his greedy instinct and attempting to produce a quick harvest of grain between successive generations of forest, he destroyed the young trees with his harrow, and with them he lost the next generation of forest. In short order this unfortunately common practice of intermixing short-term agriculture with forestry yielded both a poor rye crop and a miserable new forest.

The sight and consequences of the extensive cutting that was taking place throughout New England led Thoreau in turn to experience periods of exasperation and resignation, to develop a conservation ethic based on sound forest management, and to forge a commitment to landscape preservation. In midwinter when the citizens of Concord were accustomed to cut their forests very actively because of their freedom from other chores, the need for fuelwood, and the ease of sledding heavy loads, the extent of forest destruction seemed overwhelming to Thoreau, who emitted an exasperated "Thank God, they cannot cut down the clouds." Looking at the changes wreaked at Walden by overzealous cutting, he was forced to rationalize that ultimately the land could not be completely destroyed, for "they have done their worst and failed."

However, the strong linkage between people and the land that exists in a working landscape allowed Thoreau to develop as strong an appreciation for well-managed woodlots as for the semi-natural forest. Thus, while he delighted in the sounds, colors, and seasonal changes in the primitive woods and wild swamplands, he also admired the clean and low stumps that his acquaintance Therien left from his logging; he enjoyed the crops of berries that sprang up after a forest had been felled; and he marveled at the ability of trees like chestnut to resprout following repeated cutting. Therefore, when he considered the forested landscape that surrounded him, Thoreau clearly viewed people as an inherent, though occasionally abusive, part of the natural world. While the squirrels felled and hoarded chestnuts, the young men of Concord beat the trees with sticks and stones to bring the nuts down. These two contrasting but equally important forest activities are given even-handed treatment in Thoreau's writing. Both were inherent components of the seasonal cycle of the natural forests, and appreciation of both is essential for understanding the function of a nineteenth-century chestnut wood.

Overall, Thoreau's curiosity about natural history, his breadth of observation, and his practical inclination led him to a balanced view, one that blended ideas about forest preservation with advice on how to improve woodland practices and make them more efficient. Thoreau's personal approach to forest management was thus a broad one that involved developing a long-term plan for each woodland, based on a thorough understanding of the environment and history of the individual site, the composition of tree species, and close observation of the current abundance of regeneration by seedlings and saplings. Using this information, he would determine how the forest had developed to its current state and would begin to assess and seek to control its future development. Above

all, his approach to woodlot husbandry involved working with the natural qualities of forests and tree species rather than trying to run roughshod over nature. This outlook on forest management fits remarkably well with the sustainable approaches that are being advocated in forestry today. Thus in his critique of contemporary practices Thoreau anticipated the ecological approach to forestry that developed a century later, while also providing a detailed assessment of the consequences of the poor practices that he witnessed in his own landscape.

As he got down to specific details of "forest geometry," as he termed it, Thoreau looked at stands of different age and history and developed estimates of the longevity of species and their varying ability to establish themselves under specific conditions. He recognized that the response of a tree to a particular disturbance such as burning or cutting may differ with its size, the particular species involved, and the intensity and type of disturbance. Many species of hardwood will produce sprouts when cut, he noted, but they vary widely in the number, growth rate, and quality of the new stems produced. The abundance of tree seedlings across the landscape varies strongly with the composition of the forest, site conditions, age of the stand, and even year, since many trees produce mast crops of seeds on an irregular basis. Thoreau recognized that information from basic biology might be used to prevent the forest devastation and lack of tree regeneration that he repeatedly heard reports of from Europe. Thus he was quick to offer a series of observations and admonitions: seedling-grown trees produce better wood than sprouts; to get the best sprouts, cut the trees cleanly, low to the ground, and at a relatively small size; the sprouting ability of trees generally declines with size and age, so one should not expect a forest of large trees to be replaced by sprouts of the same species—rather there may be new tree species that seed prolifically into a forest; look carefully for seeds and seedlings in the soil or among the leaves, for these may provide the next forest; leave the slash of cut branches and limbs behind after logging because they shelter and protect young seedlings; above all, keep cattle, fire, and the plow out of a cut forest and you will reap many immediate rewards—a few years' crop of blackberries and blueberries for the townspeople, small sprouts and saplings for the rabbits, and a rapidly growing new forest that will soon yield products ranging from poles to fence posts to fuelwood.

Thoreau's recommendations for woodland practices extended beyond the individual tree and woodlot to consideration of the landscape arrangement of forests and cutting activities. On the basis of his observations of forest succession and seed dispersal by wind and animals, he argued that forests should be managed in long, narrow stands in order to enable seeds to spread from adjacent forests onto recently cut sites. This suggestion anticipates the adoption of the practice of strip clear-cutting, an approach commonly used in forests throughout the twentieth century to enhance efficient harvesting and rapid reforestation. In addition, Thoreau's description of his working with two men, a horse, and a cart to transplant 400 white pine seedlings onto two acres of land near Walden Pond is of great interest, as is the subsequent description of Emerson buying white pine seed at $4.00 per pound. Both activities are unusual indications that at least some of the citizenry of New England recognized a need to plant trees in order to establish new "artificial woods" and to counteract the regional trend of deforestation.

Thoreau's passages on woodland practices are thus an intriguing blend: commentary on the range of often destructive and counterproductive activities that he witnessed in forests across the New England countryside, and an early attempt to merge an understanding of natural history, respect for nature, and recognition of the need for forest products into a balanced approach to forest management. These journal entries are relevant to our modern landscape in conveying a sense of our forest history and providing a sound basis for our own management and conservation of natural resources.

You can walk in the woods in no direction but you hear the sound of the axe.

JANUARY 28, 1853

Even as early as 3 o'clock these winter afternoons the axes in the woods sound like nightfall, like the sound of a twilight labor.

JANUARY 8, 1852

This winter they are cutting down our woods more seriously than ever,—Fair Haven Hill, Walden, Linnaea Borealis Wood, etc., etc. Thank God, they cannot cut down the clouds!

JANUARY 21, 1852

This winter I hear the axe in almost every wood of any consequence left standing in the township.

JANUARY 9, 1855

Every larger tree which I knew and admired is being gradually culled out and carried to mill. I see one or two more large oaks in E. Hubbard's wood lying high on stumps, waiting for snow to be removed.

DECEMBER 3, 1855

These woods! Why do I not feel their being cut more sorely? Does it not affect me nearly? The axe can deprive me of much. Concord is sheared of its pride. I am certainly the less attached to my native town in consequence. One, and a main, link is broken. I shall go to Walden less frequently.

JANUARY 24, 1852

When I first paddled a boat on Walden, it was completely surrounded by thick and lofty pine and oak woods, and in some of its coves grape-vines had run over the trees and formed bowers under which a boat could pass. The hills which form its shores are so steep, and the woods on them were then so high, that, as you looked down the pond from west to east, it looked like an amphitheatre for some kind of sylvan spectacle.

JANUARY 25, 1852

The woods I walked in my youth are cut off. Is it not time that I ceased to sing? My groves are invaded.

MARCH 11, 1852

They cannot fatally injure Walden with an axe, for they have done their worst and failed.

SEPTEMBER 1, 1852

Old Mr. Joseph Hosmer, who helped me to-day, said that he used to know all about the lots, but since they've chopped off so much, and the woods have grown up, he finds himself lost.

NOVEMBER 19, 1851

Our woods are now so reduced that the chopping of this winter has been a cutting to the quick. At least we walkers feel it as such. There is hardly a wood-lot of any consequence left but the chopper's axe has been heard in it this season. They have even infringed fatally on

White Pond, on the south of Fair Haven Pond, shaved off the top-knot of the Cliffs, the Colburn farm, Beck Stow's, etc., etc.

MARCH 6, 1855

The woodchopper to-day is the same man that Homer refers to, and his work the same. He, no doubt, had his beetle and wedge and whetstone then, carried his dinner in a pail or basket, and his liquor in a bottle, and caught his woodchucks, and cut and corded, the same.

DECEMBER 29, 1853

As, on the 4th, I clambered over those great white pine masts which lay in all directions one upon another on the hillside south of Fair Haven, where the woods have been laid waste, I was struck, in favorable lights, with the jewel-like brilliancy of the sawed ends thickly bedewed with crystal drops of turpentine, thickly as a shield, as if the dryads [?], oreads [?], pine-wood nymphs had seasonably wept there the fall of the tree.

MARCH 9, 1855

We read books about logging in the Maine woods as if it were wholly strange to these parts. But I here witness almost exactly the same things, scenes that might be witnessed in Maine or New Hampshire: the logger's team, his oxen on the ice chewing the cud, the long pine tree, stripped of its branches, chained upon his sled, resting on a stout cross-bar or log and trailing behind, the smoke of his fire curling up blue amid the trees, the sound of the axe and of the teamsters' voices. A pretty forest scene, seeing oxen, so patient and stationary, good for pictures, standing on the ice,—a piece of still life. Oh, it is refreshing to see, to think of, these things after hearing of the discussions and politics of the day! The smoke I saw was quite blue. As I stood on the partially cleared bank at the east end of the pond, I looked south over the side of the hill into a deep dell still wooded, and I saw, not more than thirty rods off, a chopper at his work. I was half a dozen rods distant from the standing wood, and I saw him through a vista between two trees (it was now mainly an oak wood, the pine having been cut), and he appeared to me apparently half a mile distant, yet charmingly distinct, as in a picture of which the two trees were the frame. He was seen against the snow on the hillside beyond. I could distinguish each part of his dress perfectly, and the axe with distinct outline as he raised it above his head, the black iron against the snow, and could hear every stroke distinctly.

JANUARY 30, 1852

It is a good school the farmers' sons go to these afternoons, loading and hauling great mill-logs bigger than any cannon,—a sort of battle in the forest. I think there must be an excitement derived from their labor such as they cannot tell. After reading of the life and battles of the Northmen in Snorro Sturleson's Chronicle, these labors most remind me of that. Some of these logs are for pumps; most are for boards and timbers and spiles for bridges. I met one old pupil of mine stretched at his length upon a vast ballista, or battering-ram, of a log, while one yoke and loaded sled went on alone before and another followed behind. How they renew and wear out the paths through the woods! They think I'm loafing. I think they are drudging for gain. But no doubt our employment is more alike than we suspect, and we are each serving the great Master's ends more than our own. I have my work in the woods where I meet them, though my logs do not go to the same mill. I make a different use of skids. These men, too, who are sledding wood and sawing the logs into lengths in the woods, appear to me employed more after the old Northman fashion than the mechanics in their shops or the merchants behind their counters.

JANUARY 15, 1852

Found Therien cutting down the two largest chestnuts in the wood-lot behind where my house was [at Walden]. On the butt of one about two feet in diameter I counted seventy-five rings. T. [Therien] soon after broke his axe in cutting through a knot in this tree, which he was cutting up for posts. He broke out a piece half an inch deep. This he says often happens. Perhaps there is some frost in his axe. Several choppers have broken their axes to-day.

FEBRUARY 5, 1855

George Minott tells me that sixty years ago wood was only two or three dollars a cord here—and some of that hickory. Remembers when Peter Wheeler, sixty or more years ago, cut off all at once over a hundred acres of wood stretching from Flint's Pond to Goose Pond,—since cut again in part by Britton, and owned now partly by the Stows.

NOVEMBER 23, 1860

We do not begin to understand the treatment of woodland yet.

NOVEMBER 27, 1858

The history of a wood-lot is often, if not commonly, *here,* a history of cross-purposes,—of steady and consistent endeavor on the part of Nature, of interference and blundering with a glimmering of intelligence at the eleventh hour on the part of the proprietor. The proprietor of wood-lots commonly treats Nature as an Irishman drives a horse,—by standing before him and beating him in the face all the way across a field.

OCTOBER 16, 1860

I have come up here this afternoon to see ——'s dense white pine lot beyond the pond, that was cut off last winter, to know how the little oaks look in it. To my surprise and chagrin, I find that the fellow who calls himself its owner has burned it all over and sowed winter-rye here. He, no doubt, means to let it grow up again in a year or two, but he thought it would be clear gain if he could extract a little rye from it in the meanwhile. What a fool! Here nature had got everything ready for this emergency, and kept them ready for many years,—oaks half a dozen years old or more, with fusiform roots full charged and tops already pointing skyward, only waiting to be touched off by the sun,—and he thought he knew better, and would get a little rye out of it first, which he could feel at once between his fingers, and so he burned it, and dragged his harrow over it. As if oaks would bide *his* time or come at his bidding. Or as if he preferred to have a pine or a birch wood here possibly half a century hence—for the land is "pine sick"—rather than an oak wood at once. So he trifles with nature. I am chagrined for him. That he should call himself an agriculturalist! He needs to have a guardian placed over him. A forest-warden should be appointed by the town. Overseers of poor husband-men.

He has got his dollars for the pine timber, and now he wishes to get his bushels of grain and finger the dollars that they will bring; and then, Nature, you may have your way again. Let us purchase a mass for his soul. A greediness that defeats its own ends.

OCTOBER 16, 1860

The time will soon come, if it has not already, when we shall have to take special pains to secure and encourage the growth of white oaks, as we already must that of chestnuts for the most part. These oaks will become so scattered that there will be not seed enough to seed the ground rapidly and completely.

OCTOBER 16, 1860

You may think that you need take no care to preserve your woodland, but every tree comes either from the stump of another tree or from a seed. With the present management, will there always be a fresh stump, or a nut in the soil, think you? Will not the nobler kinds of trees, which bear comparatively few seeds, grow more and more scarce? What is become of our chestnut wood? There are but few stumps for sprouts to spring from, and, as for the chestnuts, there are not enough for the squirrels, and nobody is planting them.

DECEMBER 25, 1859

Many stumps which have only twenty-five or thirty rings send up no shoots, because they are the sprouts from old stumps, which you may still see by their sides, and so are really old trees and exhausted. The chopper should foresee this when he cuts down a wood.

NOVEMBER 25, 1860

Minott tells me that his and his sister's wood-lot together contains about ten acres and has, with a very slight exception at one time, supplied all their fuel for thirty years, and he thinks would constantly continue to do so. They keep one fire all the time, and two some of the time, and burn about eight cords in a year. He knows his wood-lot and what grows in it as well as an ordinary farmer does his corn-field, for he has cut his own wood till within two or three years; knows the history of every stump on it and the age of every sapling; knows how many beech trees and black birches there are there, as another knows his pear or cherry trees. He complains that the choppers make a very long carf [angled cut in a trunk when felling a tree] nowadays, doing most of the cutting on one side, to avoid changing hands so much. It is more economical, as well as more poetical, to have a wood-lot and cut and get out your own wood from year to year than to buy it at your door. Minott may say to his trees: "Submit to my axe. I cut your father on this very spot." How many sweet passages there must have been in his life there, chopping all alone in the short winter days! How many rabbits, partridges, foxes he saw! A rill runs through the lot, where he quenched his thirst, and several times he has laid it bare.

DECEMBER 11, 1856

We are so accustomed to see another forest spring up immediately as a matter of course, whether from the stump or from the seed, when a forest is cut down, never troubling about the succession, that we hardly associate the seed with the tree, and do not anticipate the time when this regular succession will cease and we shall be obliged to plant, as they do in all old countries. The planters of Europe must have a very different, a much correcter, notion of the value of the seed of forest trees than we. To speak generally, they know that the forest trees spring from seeds, as we do of apples and pears, but we know only that they come out of the earth.

SEPTEMBER 1, 1860

The treatment of forests is a very different question to us and to the English. There is a great difference between replanting the cleared land from the super-abundance of seed which is produced in the forest around it, which will soon be done by nature alone if we do not interfere, and the planting of land the greater part of which has been cleared for more than a thousand years.

SEPTEMBER 1, 1860

Why not control our own woods and destiny more? . . . There are many such problems in forest geometry to be solved.

OCTOBER 16, 1860

This lot is thickly covered with the rubbish or tops [the slash of branches left when the logs and firewood have been removed]. I suspect that it is, on the whole, better to leave this than to clear the ground,—that when it is not too thick (as masses of pine-tops) it is an important protection to the seedling trees (gardeners find that seedling pines require shade in their nurseries), and of course the soil is enriched by its decay.

OCTOBER 14, 1860

He [Therien] cut his trees very low, close to the ground, because the sprouts that came from such stumps were better.

DECEMBER 29, 1853

It is a pleasant surprise to walk over a hill where an old wood has recently been cut off, and, on looking round, to see, instead of dense ranks of trees almost impermeable to light, distant well-known blue mountains in the horizon and perchance a white village over an expanded open country. I now take this in preference to all my old familiar walks. So a new prospect and walks can be created where we least expected it.

NOVEMBER 9, 1850

Setting [planting] pines all day. This makes two and a half days, with two men and a horse and cart to help me. We have set some four hundred trees at fifteen feet apart diamondwise, covering some two acres. I set every one with my own hand, while another digs the holes where I indicate, and occasionally helps the other dig up the trees. We prefer bushy pines only one foot high which grow in open or pasture land, yellow-looking trees which are used to the sun, instead of the spindling dark-green ones from the shade of the woods. Our trees will not average much more than two feet in height, and we take a thick sod with them fifteen to eighteen inches in diameter. There are a great many more of these plants to be had along the edges and in the midst of any white pine wood than one would suppose. One man charged us five or six cents for them about a mile and a half distant!

APRIL 21, 1859

R. W. E. [Ralph Waldo Emerson] has bought a quarter of a pound of white pine seed at $4.00 per pound.

APRIL 21, 1859

To Walden, and set one hundred larch trees from England, all two years from seed, about nine inches high, just begun to leaf.

APRIL 29, 1859

The pickers have quite thinned the crop of early blueberries where Stow cut off [the forest] winter before last. When the woods on some hillside are cut off, the *Vaccinium Pennsylvanicum* [blueberry] springs up, or grows more luxuriantly, being exposed to light and air, and by the second year its stems are weighed to the ground with clusters of blue berries covered with bloom, and much larger than they commonly grow, also with a livelier taste than usual, as if remembering some primitive mountain-side given up to them anciently. Such places supply the villagers with the earliest berries for two or three years, or until the rising wood overgrows them and they withdraw into the bosom of Nature again. They flourish during the few years between one forest's fall and another's rise.

JULY 3, 1852

Probably the blueberry and huckleberry, amelanchier [shadbush], and other bushes which spring up immediately when the woods are cut have been already planted and started annually, as the little oaks have. Nature thus keeps a supply of these plants in her nursery (i. e. under the larger wood), always ready for casualties, as fires, windfalls, and clearings by man. Birds and foxes, etc., are annually conveying the seed into the woods.

OCTOBER 7, 1860

In the row of buttonwood [sycamore] trees on the banks of the Merrimack in Haverhill, I saw that several had been cut down, probably because of their unsightly appearance, they all suffering from the prevalent disease which has attacked the buttonwood of late years, and one large one still resting on its stump where it had fallen. It seemed like a waste of timber or of fuel, but when I inquired about it, they answered that the millers did not like to saw it. Like other ornamental trees which have stood by the roadside for a hundred years, the inhabitants have been accustomed to fasten their horses to them, and have driven many spikes into them for this purpose. One man, having carried some buttonwood logs to mill, the miller agreed to saw them if he would make good the injury which might be done to his saw. The other agreed to it, but almost at the first clip they ran on to a spike and broke the saw, and the owner of the logs cried, "Stop!" he would have no more sawed. They are difficult to split, beside, and make poor timber at best, being very liable to warp.

1850 [DAY UNKNOWN]

When the woodpeckers visit your woods in great numbers, you may suspect that it is time to cut them.

MARCH 3, 1852

I hear Barrett's sawmill running by night to improve [take advantage of] the high water. Then water is at work, another devourer of wood. These two wild forces let loose against nature. It is a hollow, galloping sound; makes tearing work, taming timber, in a rude Orphean fashion preparing it for dwellings of men and musical instruments, perchance. I can imagine the sawyer, with his lanthorn and his bar in hand, standing by, amid the shadows cast by his light. There is a sonorous vibration and ring to it, as if from the nerves of the tortured log. Tearing its entrails.

MAY 5, 1852

Went through the white cedar swamp. There are white cedars, larch (now bare), spruce, etc.; cedars two feet through, the only ones I know in Concord. It was here were cut the cedar posts which Alcott put into Emerson's summer-house. They could not be spared even for that.

NOVEMBER 14, 1853

Far the handsomest thing I saw in Boxboro was its noble oak wood. I doubt if there is a finer one in Massachusetts. Let her keep it a century longer, and men will make pilgrimages to it from all parts of the country, and yet it would be very like the rest of New England if Boxboro were ashamed of that woodland. . . . We cut down the few old oaks which witnessed the transfer of the township from the Indian to the white man, and commence our museum with a cartridge-box taken from a British soldier in 1775!

JANUARY 3, 1861

As some give to Harvard College or another institution, why might not another give a forest or huckleberry-field to Concord?

OCTOBER 15, 1859

He who cuts down woods beyond a certain limit exterminates birds.

MAY 17, 1853

Firewood and Other Fuels

In all woods is heard now far and near the sound of the woodchopper's axe, a twilight sound, now in the night of the year, men having come out for fuel to the forests,—as if men had stolen forth in the

> arctic night to get fuel to keep their fires a-going. Men go to the woods now for fuel who never go there at any other time.
>
> DECEMBER 19, 1851

In 1780 a local ordinance was passed on Block Island, Rhode Island, prohibiting the cutting of wood on common lands. In 1804 Timothy Dwight, president of Yale College, viewed the devastation brought to the land of New Hampshire by agricultural clearing and excessive logging and declared that "the forests are not only cut down but there appears little reason to hope that they will recover." In 1846 George Emerson, in his "Report on the Trees and Shrubs of Massachusetts," identified the developing shortage of forests, large trees, and fuelwood as a critical concern for the Commonwealth. From 1850 through the early 1900s, when the first state foresters were appointed in New England, wood was increasingly viewed as a natural resource that was in short supply and required careful management, government regulation, and individual attention. All this is difficult to imagine in the late twentieth century, when the forests of New England are growing in size, building products arrive in stacks of shrink-wrapped pallets from Oregon, Georgia, and Quebec, fuel is delivered in trucks or via long-distance gas lines from Texas, and an electrical grid encompasses much of the continent.

In contrast, in the mid-nineteenth century more than half of the landscape was in open agricultural land, tilled intensely to produce crops or fenced for cattle, horses, sheep, and swine. The residual forest area consisted primarily of woodlots that provided wood for building materials, fuel, packaging such as barrels, boxes, crates, and buckets, and household and industrial utensils. Large quantities of firewood were required for homes, which lacked insulation or tight-fitting windows and doors and were equipped with inefficient fireplaces, somewhat more efficient Franklin fronts, or newfangled wood stoves. Fuel was also needed in increasingly massive quantities for the new machines of the Industrial Revolution: railroads, steam engines, and turbines.

It is in the discussion of fuel and its procurement, as much as any other facet of life, that Thoreau's writing reflects the great gulf that exists between the economy and activity of his nineteenth-century landscape and our own. Although Thoreau never directly analyzed the critical reliance of New England society on fuelwood, the reality of this complete dependence on a natural resource is evident in every volume of his journal. The sounds of felling cordwood were heard throughout the land. People hoarded and collected wood scraps of every kind. Men of all backgrounds and occupations went to the forests in the fall and winter to seek their fuel. Though many people may have ignored nature on a daily basis, the citizens of nineteenth-century New England always found that they needed to turn to it annually for their vital heat.

The wealthy paid to have wood chopped and delivered, but Thoreau suggests that they actually cheated themselves by living apart from nature and not addressing their fundamental needs directly. Thoreau underlines this point by providing an amusingly critical assessment of the prosperous and successful life of his friend and mentor Ralph Waldo Emerson through an accounting of his friend's fuel economy. The annual amount of fuel needed to heat Emerson's expansive house and to support his sedentary life style of contemplation, writing, and lecturing was 25 cords of wood and approximately 14 tons of coal. In contrast, Thoreau's ne'er-do-well acquaintance John Goodwin the fisherman spent every day outside, maintained a bare shack of a residence, scavenged fuel in any inexpensive form, and burned only a cord and a half of firewood per year. Thoreau romanticized the activities involved in addressing this simple need for warmth, and he employed this as a metaphor for a life in harmony with nature. Men who seek their own wood discover that endless though heroic work is involved, as they must find, fell, chop, load, haul, split, stack, and eventually stoke it. Yet they also find that they are warmed an equal number of times and in diverse ways by this range of enterprises, as well as through the eventual combustion of wood.

Thoreau's descriptions of the diverse forms of fuel that people used in his landscape are notable, for they highlight the value and scarcity of wood and the extreme measures that were sometimes necessary in order to keep warm. Goodwin clearly was a rare opportunist. He worked the recently cleared forests and the sites of future fields for their stumps, laboriously excavating the roots by hand and cutting the resulting woody mass into transportable pieces. He also searched the riverbanks for ready-made logs and driftwood to be towed back home behind his boat. The scarcity and expense of wood are conveyed in Thoreau's use of the term "envy" when he describes such flotsam fuel. In addition, he, Goodwin, the numerous Irish immigrants, and others scrounged whatever scraps of wood they could find that would burn and yield some modicum of warmth. In one journal entry Thoreau surveyed the contents of the stove-side basket in his room in his family's house and boasted of having a distinctive and diverse assortment of fuels ready to feed his fire: chestnut fence rail, oak stump, various small limbs, white pine top, and parts of an old bridge plank. He was proud of this selection of scavenged wood and clearly relished the tales that each piece brought to his hearth.

Thoreau's descriptions of peat digging and use are equally revealing for

the light they shed on the status of woodlands and the impact of fuel procurement on the environment. The burning of the poorly decomposed peat of *Sphagnum* moss, grass, and other plants that accumulate in wetlands was a common practice in Ireland and other oceanic lands of northwestern Europe and apparently became a necessity on the island areas of Block Island, Nantucket, and Martha's Vineyard off the New England coast. Although peat has not been generally recognized as an important source of fuel in New England, Thoreau repeatedly mentions that it was harvested in Concord, referring to it as an inexhaustible resource and indicating that he considered the man who owned a peat meadow as rich indeed. He also provides some interesting details on the mechanics of peat harvesting and use. Peat, the organic matter that has accumulated over hundreds or thousands of years in lowland meadows, bogs, and swamps as a result of waterlogging and incomplete decomposition, is best cut with fork and shovel in the dry part of summer and then stacked in long, domino-like lines or open mounds to dry for weeks before being wheeled out by barrow. The result, seen through Thoreau's eyes, was a distinctive landscape scene of black, water-filled trenches running through the open wetlands, surrounded by stacked peat awaiting winter use. However, despite Thoreau's approval of peat as a fuel because of its wonderful smell, primeval connotations, and apparently inexhaustible supply, its use had a major impact on the landscape. Having taken millennia to develop, peat is actually a nonrenewable resource, and the depressions created by its removal change the surface of wetlands for decades or centuries to follow. As Thoreau notes, these depressions fill with cattails, and their excavation permanently alters the local hydrology. Once peat is mined, the landscape is never the same.

In the end, it is in his simple descriptions of the hardship of winter life and the basic need for bodily warmth that Thoreau conveys to us the intensity of nineteenth-century dependence on wood. The journal entries depicting his frigid bedroom on a January morning are graphic and unparalleled in their imagery: frost on the sheets, pails of frozen water next to the stove, spilled water shattering as ice upon contact with the floor, a frigid garret inhabited by poor neighborhood children, and drooping, blackened flowers that will never recover from their overnight freeze. This was the reality of New England life in the last century, and it underscores a major separation between Thoreau and us in our relationship with the natural environment, our forests, and our fuel. Thoreau's graphic picture of life in winter goes a long way to explain his commonly quoted statement that "each man looks upon his wood pile with a special affection," and it makes us long, as it did Thoreau, for the sultry days and mosquitoes of summer.

> There were few colder nights last winter than the last. The water in the flower-stand containing my pet tortoise froze solid,—completely enveloping him, though I had a fire in my chamber all the evening,—also that in my pail pretty thick. But the tortoise, having been thawed out on the stove, leaving the impression of his back shell in the ice, was even more lively than ever.
>
> MARCH 15, 1853

> A pail of water froze nearly half an inch thick in my chamber, with [the] fire raked up.
>
> MARCH 9, 1856

> This morning, though not so cold by a degree or two as yesterday morning, the cold has got more into the house, and the frost visits nooks never known to be visited before. The sheets are frozen about the sleeper's face; the teamster's beard is white with ice. Last night I felt it stinging cold as I came up the street at 9 o'clock; it bit my ears and face, but the stars shone all the brighter. The windows are all closed up with frost, as if they were ground glass.
>
> JANUARY 30, 1854

> The weather has considerably moderated; −2° at breakfast time (it was −8° at seven last evening); but this has been the coldest night probably. You lie with your feet or legs curled up, waiting for morning, the sheets shining with frost about your mouth. Water left by the stove is frozen thickly, and what you sprinkle in bathing falls on the floor ice. The house plants are all frozen and soon droop and turn black. I look out on the roof of a cottage covered a foot deep with snow, wondering how the poor children in its garret, with their few rags, contrive to keep their toes warm. I mark the white smoke from its chimney, whose contracted wreaths are soon dissipated in this stinging air, and think of the size of their wood-pile, and again I try to realize how they panted for a breath of cool air those sultry nights last summer. Realize it now if you can. Recall the hum of the mosquito.
>
> JANUARY 10, 1856

While Emerson sits writing [in] his study this still, overcast, moist day, Goodwin is paddling up the still, dark river. Emerson burns twenty-five cords of wood and fourteen (?) tons of coal; Goodwin perhaps a cord and a half, much of which he picks out of the river.

MARCH 18, 1857

"If I go to Boston every day and sell tape [or any merchandise] from morning till night," says the merchant (which we will admit is not a beautiful action), "some time or other I shall be able to buy the best of fuel without stint." Yes, but not the pleasure of picking it up by the riverside, which, I may say, is of more value than the warmth it yields, for it but keeps the vital heat in us that we may repeat such pleasing exercises. It warms us twice, and the first warmth is the most wholesome and memorable, compared with which the other is mere coke.

OCTOBER 22, 1853

Some farmers load their wood with gunpowder to punish thieves.

OCTOBER 12, 1855

The law requires wood to be four feet long from the middle of the carf to the middle of the carf, yet the honest deacon and farmer directs his hired men to cut his wood "four feet a little scant." He does it as naturally as he breathes.

FEBRUARY 20, 1852

I deal so much with my fuel,—what with finding it, loading it, conveying it home, sawing and splitting it,—get so many values out of it, am warmed in so many ways by it, that the heat it will yield when in the stove is of a lower temperature and a lesser value in my eyes,—though when I feel it I am reminded of all my adventures. I just turned to put on a stick. I had my choice in the box of gray chestnut rail, black and brown snag of an oak stump, dead white pine top, gray and round, with stubs of limbs, or else old bridge plank, and chose the last.

NOVEMBER 9, 1855

I pass a great white pine stump,—half a cord in it and more,—turned up out of a meadow. I look upon it with interest, and wish I had it at my door, for there are many warm fires in that. You could have many thoughts and tell many stories while that was burning.

NOVEMBER 5, 1855

I have collected and split up now quite a pile of driftwood,—rails and riders and stems and stumps of trees,—perhaps half or three quarters of a tree. It is more amusing, not only to collect this with my boat and bring [it] up from the river on my back, but to split it also, than it would be to speak to a farmer for a load of wood and to saw and split that. Each stick I deal with has a history, and I read it as I am handling it, and, last of all, I remember my adventures in getting it, while it is burning in the winter evening. That is the most interesting part of its history. It has made part of a fence or a bridge, perchance, or has been rooted out of a clearing and bears the marks of fire on it. . . . Thus one half the value of my wood is enjoyed before it is housed, and the other half is equal to the whole value of an equal quantity of the wood which I buy.

OCTOBER 20, 1855

Yesterday, toward night, gave Sophia and mother a sail as far as the Battle-Ground. One-eyed John Goodwin, the fisherman, was loading into a hand-car and conveying home the piles of driftwood which of late he had collected with his boat. It was a beautiful evening, and a clear amber sunset lit up all the eastern shores; and that man's employment, so simple and direct,—though he is regarded by most as a vicious character,—whose whole motive was so easy to fathom—thus to obtain his winter's wood,—charmed me unspeakably.

OCTOBER 22, 1853

On the 1st, when I stood on Poplar Hill, I saw a man, far off by the edge of the river, splitting billets off a stump. Suspecting who it was, I took out my glass, and beheld Goodwin, the one-eyed Ajax, in his short blue frock, short and square-bodied, as broad as for his height he can afford to be, getting his winter's wood; for this is one of the phenomena of the season. As surely as the ants which he disturbs go into winter quarters in the stump when the weather becomes cool, so does G. revisit the stumpy shores with his axe. As usual, his powder-flask peeped out from a pocket on his breast, his gun was slanted over a stump near by, and his boat lay a little further along. He had been at work laying wall still further off, and now, near the end of the day, betook himself to those pursuits which he loved better still. It would be no amusement to me to see a gentleman buy his winter wood. It is to see G. get his. I helped him tip over a stump or two. He said that the owner of the land had given him leave to get them out, but it seemed to me a condescension for him to ask any man's leave to grub up these stumps. The stumps to those who can use them, I say,—to those who will split them. He might as well ask leave of the farmer to shoot the musquash and the meadow-hen, or I might as well ask leave to look at the landscape. Near by were large hollows in the ground, now grassed over, where he had got out white oak stumps in previous years.

NOVEMBER 4, 1858

Saw Abel Brooks there with a half-bushel basket on his arm. He was picking up [wood] chips on his and neighboring lots; had got about two quarts of old and blackened pine chips, and with these was returning home at dusk more than a mile. Such a petty quantity as you would hardly have gone to the end of your yard for, and yet he said that he had got more than two cords of them at home, which he had collected thus and sometimes with a wheelbarrow. He had thus spent an hour or two and walked two or three miles in a cool November evening to pick up two quarts of pine chips scattered through the woods. He evidently takes real satisfaction in collecting his fuel, perhaps gets more heat of all kinds out of it than any man in town. He is not reduced to taking a walk for exercise as some are. It is one thing to *own* a wood-lot as he does who perambulates its bounds almost daily, so as to have worn a path about it, and another to own one as many another does who hardly knows where it is. Evidently the quantity of chips in his basket is not essential; it is the chippy idea which he pursues. It is to him an unaccountably pleasing occupation. And no doubt he loves to see his pile grow at home.

NOVEMBER 28, 1859

I see Mrs. Riordan and her little boy coming out of the woods with their bundles of fagots [firewood of small branches] on their backs. It is surprising what great bundles of wood an Irish-woman will contrive to carry. I confess that though I could carry one I should hardly think of making such a bundle of them. They are first regularly tied up, and then carried on the back by a rope,—somewhat like the Indian women and their straps. There is a strange similarity; and the little boy carries his bundle proportionally large. The sticks about four feet long. They make haste to deposit their loads before I see them, for they do not know how pleasant a sight it is to me. The Irishwoman does the squaw's part in many respects. Riordan also buys the old railroad sleepers [ties] at three dollars a hundred, but they are much decayed and full of sand.

OCTOBER 19, 1855

I saw Patrick Riordan carrying home an armful of fagots from the woods to his shanty, on his shoulder. How much more interesting an event is that man's supper who has just been forth in the snow to hunt, or perchance to steal, the fuel to cook it with! His bread and meat must be sweet.

FEBRUARY 17, 1852

Returned across Flint's Pond and the wood-lot, where some Irishman must have tried his first experiment in chopping, his first winter, where the trees were hacked off two feet from the ground, as if with a hatchet,—standing on every side of the tree by turns, and crossing the carf a hundred ways. The owner can commonly tell when an Irishman has trespassed on his wood-lot.

FEBRUARY 26, 1852

As I go down the Boston road, I see an Irishman wheeling home from far a large damp and rotten pine log for fuel. He evidently sweats at it, and pauses to rest many times. He found, perhaps, that his wood-pile was gone before the winter was, and he trusts thus to contend with the remaining cold. I see him unload it in his yard before me and then rest himself. The piles of solid oak wood which I see in other yards do not interest me at all, but this looked like *fuel.* It warmed me to think of it. He will now proceed to split it finely, and then I fear it [will] require almost as much heat to dry it, as it will give out at last.

FEBRUARY 28, 1860

I perceive that some farmers are cutting turf [grassy peat] now. They require the driest season of the year. There is something agreeable to my thoughts in thus burning a part of the earth, the stock of fuel is so inexhaustible. Nature looks not mean and niggardly, but like an ample loaf. Is not he a rich man who owns a peat meadow? It is to enjoy the luxury of wealth. It must be a luxury to sit around the fire in winter days and nights and burn these dry slices of the meadow which contain roots of all herbs. You dry and burn the very earth itself. It is a fact kindred with salt-licks. The meadow is strewn with the fresh bars, bearing the marks of the fork, and the turf-cutter is wheeling them out with his barrow. To sit and see the world aglow and try to imagine how it would seem to have it so destroyed!

AUGUST 26, 1851

I was surprised to hear Peter Flood mention it as an objection to a certain peat meadow that he would have to dry the peat on the adjacent upland. But he explained that peat dried thus was apt to crumble, and so was not so good as that dried gradually and all alike on damper ground; so an apparent disadvantage is a real advantage, according to this.

AUGUST 24, 1856

Cat-tail commonly grows in the hollows and boggy places where peat has been dug.

AUGUST 10, 1857

It will soon be forgotten, in these days of stoves, that we used to roast potatoes in the ashes, after the Indian fashion of cooking.

MARCH 2, 1852

Wildfire: A Human and Natural Force

[Fire] is without doubt an advantage on the whole. It sweeps and ventilates the forest floor, and makes it clear and clean. It is nature's besom. By destroying the punier underwood it gives prominence to the larger and sturdier trees, and makes a wood in which you can go and come. I have often remarked with how much more comfort and pleasure I could walk in woods through which a fire had run the previous year. It will clean the forest floor like a broom perfectly smooth and clear,—no twigs left to crackle underfoot, the dead and rotten wood removed,—and thus in the course of two or three years new huckleberry fields are created for the town,—for birds and men.

JUNE 21, 1850

In the course of the twentieth century Americans have undergone a major change in attitude concerning the role of fire in the natural environment. In the early 1900s a growing conviction that fire was both unnatural and highly destructive to soils, forest growth, and wildlife prompted a national effort to prevent and control the spread of wildfire. As a result of effective campaigns to raise public awareness about the damaging impact of fire, to change logging and land-use practices that contributed to accidental fires, and to develop national firefighting capabilities, a major suppression in fire activity was achieved across North America. In the coastal redwood forests and chaparral of California, in the pine and spruce forests of Yellowstone, Alaska, and the upper Midwest, and in the oak and pine forests of the East, the frequency, intensity, and extent of fires were greatly reduced. In recent decades, however, ecological studies have increasingly demonstrated that fire occurs naturally in many ecosystems whenever lightning, fuel build-up, and dry weather coincide. Meanwhile, archaeologists and anthropologists have documented that many cultures used fire for millennia to manage their landscapes. Consequently, in many grassland, shrubland, and forested areas the plants have adapted to survive fire, and in fact the entire ecosystem may actually require a regular regime of burning in order to maintain its composition and natural processes through time. The frequency, intensity, and seasonality of fire are often critical factors that control the basic structure, plant composition, and movement of energy and nutrients in these ecosystems. Consequently, if the established fire regime is modified through effective suppression activities, the whole ecosystem may change.

One result of this emerging information about the natural role and cultural use of fire is that scientists and land managers have begun to appreciate fire as an important ecological process and to employ it as a useful management tool. In some large natural areas and national parks, lightning fires have been allowed to burn under supervision, while in other landscapes prescribed fires have been set purposely and then closely controlled in order to achieve specific management objectives, including the maintenance and enhancement of natural processes and characteristics.

The historical role of fire in New England, either as a natural process or as a tool used by Indians, remains a major enigma for archaeologists and ecologists; the records are fragmentary, and interpretations necessarily involve considerable speculation. Frequent or extensive lightning fires are considered to be rare in this relatively moist region, and therefore natural fire is perceived as generally unimportant. However, Indians are known to have used fire around village sites, and some ecologists and archaeologists believe that Indians were a major source of ignition across the broader countryside. In his reconstructions of New England forest history, Thoreau cites William Wood's seventeenth-century descriptions of the prolific use of fire by Indians to modify the natural vegetation and environment. Wood and other early explorers and colonists, including Thomas Morton, suggested that the Indians set fires to clear the forested and brushy areas around villages and to open the understory of woods to improve travel and wildlife habitats. The historical occurrence of Indian-set fires is therefore an established fact, but the extent and frequency of the practice still remain unclear.

Similarly, we know relatively little about the incidence, cause, and types of fire that occurred through the colonial period and into the end of the last century. Reasonable speculation, based on some limited historical and paleoecological data, suggests that fire increased with the growth in human population and the widespread land-clearing activities that occurred during the nineteenth century. For this period, Thoreau's journals provide a great deal of insight and information. It is apparent from his observations that fire was startlingly common. In the spring and autumn the sky was often dominated by a smoky haze, and columns of smoke could be seen on the horizon and from many hilltops; thus the evidence for widespread fire was apparent in both the air and the woods. Through Thoreau's descriptions, we can learn more about the cultural role of fire in New England and can begin to search for the legacy of this disturbance in the modern landscape.

Thoreau's evaluations of the causes, seasonal timing, and impact of fire in New England are intriguing. Although he mentions lightning as a potential source of wildfire, he immediately dismisses it as an insignificant factor in his landscape. Rather, three specific causes for the ignition of wildfires, all of human origin, emerge in his writings: locomotives, the wadding from hunters' rifles, and the burning of brush and meadows by farmers. Each of these sources was capable of producing widely spreading fires that could burn extensive areas, and each had its particular distribution in the landscape. In the days before spark-arresting screens were placed on locomotive smokestacks to capture flying embers, railroads were a major source of ignitions. Fires that started along train tracks spread down the embankments and into the adjacent fields and forests and were a constant source of problems for the citizens of rural towns. Thoreau's description of the smoke-belching, spark-spewing passage of trains along the Fitchburg Railroad makes this concern quite understandable.

Thoreau does not linger on the subject of fires that originated from gun

wadding, although it is an interesting one. With the style of rifle used in his time, a spark-laden wad was frequently ejected with the shot. If unnoticed by a hunter, the small coals on this cotton tinder could gradually build into a blaze. Since hunters roamed parts of the landscape that were distant from the railroad, field edge, and town center, gun wadding was a source of ignition that could generate a distinctive pattern of fire throughout the older woodlands and less frequented areas. Because of their relative remoteness, these fires were often extensive and difficult to control.

The annual burning of fields following the harvest of grain and hay, or the brush burning that followed woodland clearance, provided the major source of the numerous columns of smoke that Thoreau saw spreading across the horizon. The widespread burning of fields may be difficult to imagine in New England today, but it has remained a common practice in many grain and grassland areas in the midwestern and western United States and across northwestern Europe. Fire removed the stubble of grass and matted hay from the fields and, according to Thoreau, was followed rapidly by a new growth of green shoots. In the meadows, however, Thoreau noted that the fire could also burn deeply into the underlying peat where it might persist for days or weeks, excavating large holes and giving the impression of geothermal heat emanating from the earth.

Overall, Thoreau's evaluation of the ecological effect of fire is quite positive, as befits a person living in a landscape where fire was a commonly employed process that had a regular impact on forests. Although extremely intense fire can kill entire forests and remove the surface layers of the soil, the general impact of fire is beneficial. Thoreau suggests that burning cleans the forest floor of debris and selectively removes the smaller, less vigorous plant growth in favor of larger trees. The result is a great improvement: an open forest good for walking that is composed of more productive trees. Plants respond dramatically to fire as it changes the surrounding environment; they often grow much faster because of the release of nutrients in the ash and the decreased competition for light and water from other plants that have been killed. Thick-barked, fire-resistant tree species such as oaks and pines are generally only blackened by surface fires, although their lower limbs may die. Many of the hardwood tree species that grow in New England sprout prolifically after injury by fire, including the oaks, maples, cherries, birches, chestnuts, and hickories.

Among the trees, Thoreau's favorite, the pitch pine, exhibits the most remarkable behavior with regard to fire, for it is the only northern conifer that is capable of sprouting. After fire or damage by cutting, pitch pine can produce new shoots, branches, and needles from small buds that are shielded beneath the bark at the base of the tree and along the length of the trunk and branches. Consequently, although fires often roar through a highly flammable stand of pitch pines, burning the dry needles on the ground and consuming the green needles on the living shoot to leave blackened, leafless stems, the trees frequently survive. Within weeks, new green growth appears across the stems as bundles of fresh needles emerge from the buried buds. Phoenix-like, the intensely scorched pine recovers and grows anew. Most of the shrubs and herbs in our forests are equally resilient because of their extensive network of roots and underground rhizomes, which are protected in the soil from the intense heat that is released above ground. All of the surviving species can grow rapidly with the increase in resources, and most exhibit an increased production of flowers and fruits in the aftermath of a severe fire. Thus, Thoreau notes the abundance of huckleberry and blueberries on burned land and identifies the common herbs, the bellwort (uvularia), false Solomon's seal (smilacina), and grasses, as young green shoots among the charcoal on the ground.

The seasonal timing of fire, which is controlled by the availability of ignition, fuel, and the weather, is critical in determining its impact. In New England's largely deciduous forest, the green growth of foliage throughout the midsummer months is a great impediment to fire because its high moisture content precludes combustion and its shade keeps the forest understory cool, moist, and relatively windless. Consequently, Thoreau notes, most fires occur in the spring, after the frost and moisture have left the ground and before the leaves have greened out. The other major season for New England fires is fall, when the leaves are dropping and farmers burn fields that have been harvested. During these two periods, and especially following a drought, Thoreau frequently observed fires of considerable size and intensity.

To Thoreau fire was both a common part of the landscape and a general benefit to natural and agricultural ecosystems. His descriptions make it clear that fire was important to the livelihood of the people and was a strong force in shaping the composition and dynamics of the forests, shrublands, and fields that form our landscape today. An understanding of this historical role of fire in shaping the characteristics of the New England countryside is valuable to us as we seek to manage the modern landscape. In particular, Thoreau's descriptions of the seasonality, intensity, and impact of fires can provide some guidance to land managers aiming to recreate the conditions that gave rise to particular types of vegetation or attempting to maintain populations of plant or animal species that may be dependent on fire and other disturbances for their continuity in our countryside.

The fires in woods and meadows have been remarkably numerous and extensive all over the country, the earth and vegetation have been so dry, especially along railroads and on mountains and pine plains. Some meadows are said to have been burned three feet deep. On some mountains it burns all the soil down to the rock. It catches from the locomotive, from sportsmen's [rifle] wadding, and from burning brush and peat meadows. In all villages they smell smoke, especially at night. On Lake Champlain, the pilots of steamboats could hardly see their course, and many complained that the smoke made their eyes smart and affected their throats.

SEPTEMBER 2, 1854

Already I hear of a small fire in the woods in Emerson's lot, set by the [locomotive] engine, the leaves that are bare are so dry.

APRIL 1, 1856

The ground is so completely bare this winter, and therefore the leaves in the woods so dry, that on the 5th there was a fire in the woods by Walden (Wheeler's), and two or three acres were burned over, set probably by the engine. Such a burning as commonly occurs in the spring.

FEBRUARY 8, 1858

The woods are in a state of tinder, and the smoker and sportsman and the burner must be careful now.

MARCH 29, 1858

I hear that there has been a great fire in the woods this afternoon near the factory. Some say a thousand acres have been burned over.

MARCH 31, 1860

Now commences the season for fires in the woods. The winter, and now the sun and winds, have dried the old leaves more thoroughly than ever, and there are no green leaves to shade the ground or to check the flames, and these high March winds are the very ones to spread them . . . With these thoughts and impressions I had not gone far before I saw the smoke of a fire on Fair Haven Hill. Some boys were going *sassafrasing* [collecting sassafras], for boys will have some pursuit peculiar to every season. A match came in contact with a marble, nobody knew how, and suddenly the fire flashed up the broad open hillside, consuming the low grass and sweet-fern and leaving a smoking, blackened waste. A few glowing stumps, with spadefuls of fresh earth thrown on them, the white ashes here and there on the black ground, and the not disagreeable scent of smoke and cinders was all that was left when I arrived.

MARCH 30, 1853

Look out for your wood-lots between the time when the dust first begins to blow in the streets and the leaves are partly grown.

APRIL 2, 1860

I smelled the burnt ground a quarter of a mile off. It was a very severe burn, the ground as black as a chimney-back. The fire is said to have begun by an Irishman burning brush near Wild's house in the south part of Acton, and ran north and northeast some two miles before the southwest wind, crossing Fort Pond Brook. I walked more than a mile along it and could not see to either end, and crossed it in two places. A thousand acres must have been burned. The leaves being thus cleanly burned, you see amid their cinders countless mouse-galleries, where they have run all over the wood, especially in shrub oak land, these lines crossing each other every foot and at every angle. You are surprised to see by these traces how many of these creatures live and run under the leaves in the woods, out of the way of cold and of hawks. The fire has burned off the top and half-way down their galleries. Every now and then we saw an oblong square mark of pale-brown or fawn-colored ashes amid the black cinders, where corded [fire] wood had been burned.

This fire ran before the wind, which was southwest, and, as nearly as I remember, the fires generally at this season begin on that side, and you need to be well protected there by a plowing or raking away the leaves. Also the men should run ahead of the fire before the wind, most of them, and stop it at some cross-road, by raking away the leaves and setting back fires.

APRIL 2, 1860

It is to be observed that we heard of fires in the woods in various towns, and more or less distant, on the same days that they occurred here,—the last of March and first of April. The newspapers reported many. The same cause everywhere produced the same effect.

APRIL 5, 1860

Surveying Emerson's wood-lot to see how much was burned near the end of March, I find that what I anticipated is exactly true,—that the fire did not burn hard on the northern slopes, there being then frost in the ground, and where the bank was very steep, say at angle of forty-five degrees, which was the case with more than a quarter of an acre, it did not run down at all, though no man hindered it.

APRIL 30, 1860

Here, by the side of the pond, a fire has recently run through the young woods on the hillside. It is surprising how clean it has swept the ground, only the very lowest and dampest rotten leaves remaining, but uvularias [bellworts] and smilacinas [false Solomon's seal] have pushed up here and there conspicuously on the black ground, a foot high. At first you do not observe the full effect of the fire, walking amid the bare dead or dying trees, which wear a perfect winter aspect, which, as trees generally are not yet fully leaved out and you are still used to this, you do not notice, till you look up and see the still green tops everywhere above the height of fifteen feet. Yet the trees do not bear many marks of fire commonly; they are but little blackened except where the fire has run a few feet up a birch, or paused at a dry stump, or a young evergreen has been killed and reddened by it and is now dropping a shower of red leaves.

MAY 20, 1853

As I go home by Hayden's I smell the burning meadow. I love the scent. It is my pipe. I smoke the earth.

AUGUST 14, 1854

At this season, too, the farmers burn brush, and the smoke is added to the haziness of the atmosphere. From this hill I count five or six smokes, far and near, and am advertised of one species of industry over a wide extent of country.

AUGUST 21, 1852

Crossed the Brooks or Hadlock meadows, which have been on fire (spread from bogging [digging and burning of peat, roots and shrubs to improve the meadow land]) several weeks. They present a singularly desolate appearance. Much of the time over shoes in ashes and cinders. Yellowish peat ashes in spots here and there. The peat beneath still burning, as far as dry, making holes sometimes two feet deep, they say. The surface strewn with cranberries burnt to a cinder. I seemed to feel a dry heat under feet, as if the ground were on fire, where it was not.

AUGUST 23, 1854

Am surprised to see how green the forest floor and the sprout-land north of Damon's lot are already again, though it was a very severe burn. In the wood-lot the trees are *apparently* killed for twenty feet up, especially the smaller, then six or ten feet of green top, while very vigorous sprouts have shot up from the base below the influence of the fire. This shows that they will die, I think. The top has merely lived for the season while the growth has been in their sprouts around the base. This is the case with oaks, maples, cherry, etc. Also the blueberry *(Vaccinium vacillans)* has sent up very abundant and vigorous shoots all over the wood from the now more open and cleaned ground. These are evidently from stocks [shoots and roots] which were comparatively puny before. The adjacent oak sprout-land has already sprung up so high that it makes on me about the same impression that it did before, though it [was] from six to ten feet high and was generally killed to the ground. . . . Thus the stumps and roots of young oak, chestnut, hickory, maple, and many other trees retain their vitality a very long time and after many accidents, and produce thrifty trees at last.

OCTOBER 8, 1860

And then, of course, there was the day when Thoreau, along with his companion Edward Hoar, started what became the most infamous fire in the history of Concord. Spreading from their campfire and fanned by breezes following a period of lengthy drought, this blaze burned an extensive area of field and woods and sealed Thoreau's reputation as a lazy scallywag in the minds of many of his fellow citizens. Grumbling about the blackened land and the enterprises of careless ne'er-do-wells followed him for years through the village streets. Nonetheless, Thoreau remained defiant, regarding this event in a positive light; it has come to be one of the most frequently cited stories about this American naturalist, his ability to rationalize his actions, and his relationship to the environment and his contemporaries.

> I once set fire to the woods. Having set out, one April day, to go to the sources of Concord River in a boat with a single companion, meaning to camp on the bank at night or seek a lodging in some neighboring country inn or farmhouse, we took fishing tackle with us that we might fitly procure our food from the stream, Indian-like. At the shoemaker's near the river, we obtained a match, which we had forgotten. Though it was thus early in the spring, the river was low, for there had not been much rain, and we succeeded in catching a mess of fish sufficient for our dinner before we had left the town, and by the shores of Fair Haven Pond we proceeded to cook them. The earth was uncommonly dry, and our fire, kindled far from the woods in a sunny recess in the hillside on the east of the pond, suddenly caught the dry grass of the previous year which grew about the stump on which it was kindled. We sprang to extinguish it at first with our hands and feet, and then we fought it with a board obtained from the boat, but in a few minutes it was beyond our reach; being on the side of a hill, it spread rapidly upward, through the long, dry, wiry grass interspersed with bushes.
>
> "Well, where will this end?" asked my companion. I saw that it might be bounded by Well Meadow Brook on one side, but would, perchance, go to the village side of the brook. "It will go to town," I answered. While my companion took the boat back down the river, I set out through the woods to inform the owners and to raise the town. The fire had already spread a dozen rods on every side and went leaping and crackling wildly and irreclaimably toward the wood. That way went the flames with wild delight, and we felt that we had no control over the demonic creature to which we had given birth. We had kindled many fires in the woods before, burning a clear space in the grass, without ever kindling such a fire as this.
>
> As I ran toward the town through the woods, I could see the smoke over the woods behind me marking the spot and the progress of the flames. The first farmer whom I met driving a team, after leaving the woods, inquired the cause of the smoke. I told him. "Well," said he, "it is none of my stuff," and drove along. The next I met was the owner in his field, with whom I returned at once to the woods, running all the way. I had already run two miles. When at length we got into the neighborhood of the flames, we met a carpenter who had

been hewing timber, an infirm man who had been driven off by the fire, fleeing with his axe. The farmer returned to hasten more assistance. I, who was spent with running, remained. What could I do alone against a front of flame half a mile wide?

I walked slowly through the wood to Fair Haven Cliff, climbed to the highest rock, and sat down upon it to observe the progress of the flames, which were rapidly approaching me, now about a mile distant from the spot where the fire was kindled. Presently I heard the sound of the distant bell giving the alarm, and I knew that the town was on its way to the scene. Hitherto I had felt like a guilty person,—nothing but shame and regret. But now I settled the matter with myself shortly. I said to myself: "Who are these men who are said to be the owners of these woods, and how am I related to them? I have set fire to the forest, but I have done no wrong therein, and now it is as if the lightning had done it. These flames are but consuming their natural food." (It has never troubled me from that day to this more than if the lightning had done it. The trivial fishing was all that disturbed me and disturbs me still.) So shortly I settled it with myself and stood to watch the approaching flames. It was a glorious spectacle, and I was the only one there to enjoy it. The fire now reached the base of the cliff and then rushed up its sides. The squirrels ran before it in blind haste, and three pigeons dashed into the midst of the smoke. The flames flashed up the pines to their tops, as if they were powder.

When I found I was about to be surrounded by the fire, I retreated and joined the forces now arriving from the town. It took us several hours to surround the flames with our hoes and shovels and by back fires subdue them. In the midst of all I saw the farmer whom I first met, who had turned indifferently away saying it was none of his stuff, striving earnestly to save his corded wood, his stuff, which the fire had already seized and which it after all consumed.

It burned over a hundred acres or more and destroyed much young wood. When I returned home late in the day, with others of my townsmen, I could not help noticing that the crowd who were so ready to condemn the individual who had kindled the fire did not sympathize with the owners of the wood, but were in fact highly elate[d?] and as it were thankful for the opportunity which had afforded them so much sport; and it was only half a dozen owners, so called, though not all of them, who looked sour or grieved, and I felt that I had a deeper interest in the woods, knew them better and should feel their loss more, than any or all of them. The farmer whom I had first conducted to the woods was obliged to ask me the shortest way back, through his own lot. Why, then, should the half-dozen owners [and] the individuals who set the fire alone feel sorrow for the loss of the wood, while the rest of the town have their spirits raised? Some of the owners, however, bore their loss like men, but other some declared behind my back that I was a "damned rascal;" and a flibbertigibbet or two, who crowed like the old cock, shouted some reminiscences of "burnt woods" from safe recesses for some years after. I have had nothing to say to any of them. The locomotive engine has since burned over nearly all the same ground and more, and in some measure blotted out the memory of the previous fire. For a long time after I had learned this lesson I marvelled that while matches and tinder were contemporaries the world was not consumed; why the houses that have hearths were not burned before another day; if the flames were not as hungry now as when I waked them. I at once ceased to regard the owners and my own fault,—if fault there was any in the matter,—and attended to the phenomenon before me, determined to make the most of it. To be sure, I felt a little ashamed when I reflected on what a trivial occasion this had happened, that at the time I was no better employed than my townsmen.

That night I watched the fire, where some stumps still flamed at midnight in the midst of the blackened waste, wandering through the woods by myself; and far in the night I threaded my way to the spot where the fire had taken, and discovered the now broiled fish,—which had been dressed,—scattered over the burnt grass.

1850 [DAY UNKNOWN]

When the lightning burns the forest its Director makes no apology to man, and I was but His agent. Perhaps we owe to this accident partly some of the noblest natural parks. It is inspiriting to walk amid the fresh green sprouts of grass and shrubbery pushing upward through the charred surface with more vigorous growth.

JUNE 21, 1850

The Coming of the New Forest

Social Change and Farm Abandonment in New England

> As for antiquities, one of our old deserted country roads, marked only by the parallel fences and cellar-hole with its bricks where the last inhabitant died, the victim of intemperance, fifty years ago, with its bare and exhausted fields stretching around, suggests to me an antiquity greater and more remote from the America of the newspapers than the tombs of Etruria. . . . This is the decline and fall of the Roman Empire.
>
> FEBRUARY 13, 1851

By the time Thoreau died in 1862, the balance had already tipped in New England and the astonishing march of forest across the open landscape had begun. Driven by the individual decisions of tens of thousands of farmers and their families across the eastern United States and played out slowly on countless hillslopes and pastures, the process of gradual abandonment of agriculture and the natural invasion of old fields by native trees, shrubs, and herbs was under way. Today, we accept and live with the consequences of this transformation, giving it little thought. We may observe a stone wall in the forest and conclude that farming was not productive in the rocky hills of New England, that futile decades of hardscrabble existence came to an end when the inhabitants moved to more productive lands farther west. More likely we do not even recognize the stone wall for what it is, nor do we try to understand or explain the process that gradually but irrevocably transformed New England life and landscape more than a century ago.

Many questions come to mind. Why did the prosperous farmers, along with the not so prosperous ones, decide to abandon lands and a way of life that had occupied generations of New Englanders? After two centuries of backbreaking work clearing and shaping the soil into an orderly and productive agricultural landscape, what could have persuaded the rural occupants of New England to abandon their farms, to leave their churches, friends, and towns, and to allow the land to return piecemeal to forest? How was this change in the people and the land perceived at the time, and what feelings did it evoke? Where were the people drawn as they uprooted themselves and moved their lives from the New England soil? And how did the natural process of forest development play out in the increasingly abandoned fields and meadows and woodlots? The official U.S. census may record the stark statistics of this major change in New England, but Henry Thoreau quietly captured its spirit.

Thoreau's writings on agriculture convey a surprising sense of spirited energy, vitality, and progress that does not support the stereotype of hard-scrabble rural life in the northeastern United States. Thoreau's passages are filled with descriptions of the abundance of crops, the productivity of the land, and the general success of its inhabitants; from his vantage point farming in New England was a highly profitable, though endlessly back-breaking, enterprise. True, there were unwise and lazy farmers, eroded hillslopes, and less productive farms, but seldom in reading Thoreau's journals does one receive the impression that the natural quality of the land was a limiting factor in agriculture, that the soil was too miserable or too poor to yield productive harvests, or that farmers ever began to exhaust the potential of the land. In fact, quite the opposite appears true. By amending soils with manure, muck, and lime, farmers were continually enhancing the natural fertility of the land. By ditching and draining wetlands and by clear-cutting forests, they progressively expanded the area of productive farmland. And by continually removing stones from the ground, they improved the ease of plowing year by year. Every indication from Thoreau's writing, as well as from other historical sources, suggests that agricultural yields from New England actually increased through the middle part of the century and that farming provided a productive and well-respected life style. The homes built throughout the rural landscape and in village centers by Thoreau's contemporaries remain as grand and substantial testimony to their owners' prosperity and their anticipation of a long life in New England.

What happened to drive the farm families from their land and livelihoods? Thoreau's journal identifies a few of the key factors, only one of which directly involved the practice of farming itself. The major factor was a social transformation in America, a cultural revolution based on industrialization, which generated a widespread discontent with the old rural life style, an interest and growing excitement in change, and an opportunity to join the activity and apparent economic well-being of city life. The railroad that Thoreau simultaneously detested and romanticized in his writings was a striking image of this change. Trains provided rapid travel and connections with other places, and they exposed the rural population to the goods and fashions of distant locations. The railroad hinted of cities and of foreign sights and helped to initiate a generational shift in attitude. Sons lost the desire to farm. If they could not be in California or the cities, they and their young wives wanted to be near the rail line so that they could feel connected to these exciting new possibilities. In the space of one generation, starting around the middle of the nineteenth century, New England changed. Thoreau became aware of a new mood: the rural landscape was perceived as "seed country," far from the post office, train, and telegraph. For many people, as Thoreau poignantly described it, the countryside began to feel lonely: he mentions a woman who was unable to bear the solitude, a man who sought out his horse for added companionship.

There were two major attractions that led to this malaise, according to Thoreau: California, which provided new, though transient, opportunities in its recently discovered gold fields, and eastern cities, which offered jobs, market goods, and a new life style. It is evident what Thoreau, the man who found all of life inside himself, thought of both. To him, the gold rush was a lottery where soulless men would seek in vain to become rich without work. The cities and their factories were lifeless; the employment they offered was debasing to the human spirit and the antithesis of the heroic life of the countryside. From his travels by boat down the Merrimack River through Lowell, by train through Fitchburg, Gardner, and Nashua, and by foot across Staten Island, Thoreau had received firsthand views of the urban life and factories of the Industrial Revolution. His description of the vast interior of the gingham mill at Clinton, Massachusetts, captures the starkness, immensity, and impersonal qualities of the industrial age that appear in contemporary etchings and photographs. The influence that the factories and dams exerted on nature by constricting and controlling the flow of the rivers provided Thoreau with a metaphor for the increasing impact they had on man's relationship to nature in his own lifetime.

The railroad also had a direct impact on agriculture. The trains brought farm produce, industrial products, and unanticipated competition. The farmers of New England began to use grain from the southern and western part of the country at their own tables, while restricting the use of local corn and grain to their livestock. The availability of new products provided intense competition for local industry and small farms as well as creating a great longing for material possessions. Both developments irked Thoreau, who wrote that bread stuffs arrived from the west, butter stuffs from Vermont, and other items that "we stuff ourselves" with came from around the globe as the distribution of agricultural and industrial production expanded. New England agriculture increasingly specialized in a narrow range of products that could not be transported easily, such as milk and vegetables, and this increased the need for hard cash to purchase everything else.

Interestingly, Thoreau's journals indicate that the abandonment of fields, which would become so prevalent from the 1850s onward, was actually occurring at a regular pace throughout the century. Evidently, farming always represented a certain dynamic balance between the creation and expansion of new fields and the abandonment of existing ones. In 1850, for instance, Thoreau described a stone wall running through an old wood in Concord and remarked on the stability of these seemingly fragile fences. The old pasture that this forest grew up in must have been abandoned in the early 1800s.

It is clear, however, that Thoreau sensed a real and perceptible change in the regional agricultural balance by mid-century. In 1850 he could walk deserted country roads marked only by cellar holes and crumbling fences; he could follow abandoned cowpaths crowded with dense young birches as he tried to relocate the old apple orchards that they had engulfed; and he could recognize a major shift in his friend Hosmer's neighborhood as all of the old farms were bought up by one individual, and the blacksmith, goldsmith, tavern keeper, and store owner packed up, leaving a deserted country. To Thoreau, these scenes suggested that the vitality of America was in decline; the New World suddenly seemed old.

> The railroads as much as anything appear to have unsettled the farmers. Our young Concord farmers and their young wives, hearing this bustle about them, seeing the world all going by as it were,— some daily to the cities about their business, some to California,— plainly cannot make up their minds to live the quiet, retired,

old-fashioned, country-farmer's life. They are impatient if they live more than a mile from a railroad. While all their neighbors are rushing to the road, there are few who have character or bravery enough to live off the road. He is too well aware what is going on in the world not to wish to take some part in it.

SEPTEMBER 28, 1851

As I stand under the hill beyond J. Hosmer's and look over the plains westward toward Acton and see the farmhouses nearly half a mile apart, few and solitary, in these great fields between these stretching woods, out of the world, where the children have to go far to school; the still, stagnant, heart-eating, life-ever-lasting, and gone-to-seed country, so far from the post-office where the weekly paper comes, wherein the new-married wife cannot live for loneliness, and the young man has to depend upon his horse for society; see young J. Hosmer's house, whither he returns with his wife in despair after living in the city,—I standing in Tarbell's road, which he alone cannot break out [i.e., maintain a good open road during the snowy winter months],—the world in winter for most walkers reduced to a sled track winding far through the drifts, all springs sealed up and no digressions; where the old man thinks he may possibly afford to rust it out, not having long to live, but the young man pines to get nearer the post-office and the Lyceum, is restless and resolves to go to California, because the depot is a mile off (he hears the rattle of the cars at a distance and thinks the world is going by and leaving him); where rabbits and partridges multiply, and muskrats are more numerous than ever, and none of the farmer's sons are willing to be farmers, and the apple trees are decayed, and the cellar-holes are more numerous than the houses, and the rails are covered with lichens, and the old maids wish to sell out and move into the village, and have waited twenty years in vain for this purpose and never finished but one room in the house, never plastered nor painted, inside or out, lands which the Indian was long since dispossessed [of], and now the farms are run out, and what were forests are grain-fields, what were grain-fields, pastures . . . I say, standing there and seeing these things, I cannot realize that this is that hopeful young America which is famous throughout the world for its activity and enterprise, and this is the most thickly settled and Yankee part of it. What must be the condition of the *old* world!

JANUARY 27, 1852

The recent rush to California and the attitude of the world, even of its philosophers and prophets, in relation to it appears to me to reflect the greatest disgrace on mankind. That so many are ready to get their living by the lottery of gold-digging without contributing any value to society, and that the great majority who stay at home justify them in this both by precept and example! It matches the infatuation of the Hindoos who have cast themselves under the car of Juggernaut. I know of no more startling development of the morality of trade and all the modes of getting a living than the rush to California affords. Of what significance the philosophy, or poetry, or religion of a world that will rush to the lottery of California gold-digging on the receipt of the first news, to live by luck, to get the means of commanding the labor of others less lucky, *i. e.* of slaveholding, without contributing any value to society? And that is called enterprise, and the devil is only a little more enterprising! The philosophy and poetry and religion of such a mankind are not worth the dust of a puffball. The hog that *roots* his own living, and so makes manure, would be ashamed of such company. If I could command the wealth of all the worlds by lifting my finger, I would not pay such a price for it. It makes God to be a moneyed gentleman who scatters a handful of pennies in order to see mankind scramble for them. Going to California. It is only three thousand miles nearer to hell. I will resign my life sooner than live by luck. The world's raffle. A subsistence in the domains of nature a thing to be raffled for! No wonder that they gamble there. I never heard that they did anything else there. What a comment, what a satire, on our institutions! The conclusion will be that mankind will hang itself upon a tree. And who would interfere to cut it down. And have all the precepts in all the bibles taught men only this? and is the last and most admirable invention of the Yankee race only an improved muck-rake?—patented too! If one came hither to sell lottery tickets, bringing satisfactory credentials, and the prizes were seats in heaven, this world would buy them with a rush.

Did God direct us so to get our living, digging where we never planted,—and He would perchance reward us with lumps of gold? It is a text, oh! for the Jonahs of this generation, and yet the pulpits are as silent as immortal Greece [?], silent, some of them, because the preacher is gone to California himself. The gold of California is a touchstone which has betrayed the rottenness, the baseness, of mankind. Satan, from one of his elevations, showed mankind the kingdom of California, and they entered into a compact with him at once.

FEBRUARY 1, 1852

Men rush to California and Australia as if the true gold were to be found in that direction; but that is to go to the very opposite extreme to where it lies. They go prospecting further and further away from the true lead, and are most unfortunate when most successful. Is not our native soil auriferous?

OCTOBER 18, 1855

Here we are, deriving our breadstuffs from the West, our butter stuffs from Vermont, and our tea and coffee and sugar stuffs, and much more with which we stuff ourselves, from the other side of the globe.

OCTOBER 17, 1859

The whistle of the locomotive penetrates my woods summer and winter, sounding like the scream of a hawk sailing over some farmer's yard, informing me that many restless city merchants are arriving within the circle of the town, or adventurous country traders from the other side. As they come under one horizon, they shout their warning to get off the track to the other, heard sometimes through the circles of two towns. Here come your groceries, country; your rations, countrymen! Nor is there any man so independent on his farm that he can say them nay. And here's your pay for them! screams the countryman's whistle; timber like long battering-rams going twenty miles an hour against the city's walls, and chairs enough to seat all the weary and heavy-laden that dwell within them. With such huge and lumbering civility the country hands a chair to the city. All the Indian huckleberry hills are stripped, all the cranberry meadows are raked into the city. Up comes the cotton, down goes the woven cloth; up comes the silk, down goes the woolen; up come the books, but down goes the wit that writes them.

WALDEN, p. 82

The Boiling Spring [a familiar source of good water in Concord] is turned into a tank for the Iron Horse [i.e., the locomotive] to drink at, and the Walden woods have been cut and dried for his fodder. That devilish Iron Horse, whose earrending whinner is heard throughout the town, has defiled the Boiling Spring with his feet and drunk it up, and browsed off all the wood around the pond. He has got a taste for berries even, and with unnatural appetite he robs the country babies of milk, with the breath of his nostrils polluting the air. That Trojan horse, with a thousand men in his belly, insidiously introduced by mercenary Greeks. With the scream of a hawk he beats the bush for men, the man-harrier, and carries them to his infernal home by thousands for his progeny. Where is the country's champion, the Moore of Moore Hall, to meet him at the Deep Cut and throw a victorious and avenging lance against this bloated pest?

JUNE 17, 1853

I find three or four ordinary laborers to-day putting up the necessary outdoor fixtures for a magnetic telegraph from Boston to Burlington. They carry along a basket full of simple implements, like travelling tinkers, and, with a little rude soldering, and twisting, and straightening of wires, the work is done. It is a work which seems to admit of the greatest latitude of ignorance and bungling, and as if you might set your hired man with the poorest head and hands to building a magnetic telegraph. All great inventions stoop thus low to succeed, for the understanding is but little above the feet.

AUGUST 28, 1851

But now, by means of railroads and steamboats and telegraphs, the country is denaturalized. . . .

FEBRUARY 8, 1854

It is surprising what a tissue of trifles and crudities make the daily news. For one event of interest there are nine hundred and ninety-nine insignificant, but about the same stress is laid on the last as on the first. The newspapers have just told me that the transatlantic telegraph-cable is laid. That is important, but they instantly proceed to inform me how the news was received in every larger town in the United States,—how many guns they fired, or how high they jumped,—in New York, and Milwaukee, and Sheboygan; and the boys and girls, old and young, at the corners of the streets are reading it all with glistening eyes, down to the very last scrap, not omitting what they did at New Rochelle and Evansville. And all the speeches are reported, and some think of collecting them into a volume!!!

AUGUST 9, 1858

Hear by telegraph that it rains in Portland and New York.

AUGUST 26, 1854

Saw at Clinton last night a room at the gingham-mills which covers one and seven-eighths acres and contains 578 looms, not to speak of spindles, both throttle and mule. The rooms all together cover three acres. They were using between three and four hundred horse-power, and kept an engine of two hundred horse-power, with a wheel twenty-three feet in diameter and a band ready to supply deficiencies, which have not often occurred. Some portion of the machinery—I think it was where the cotton was broken up, lightened up, and mixed before being matted together—revolved eighteen hundred times in a minute.

JANUARY 2, 1851

We admire more the man who can use an axe or adze skillfully than him who can merely tend a machine. When labor is reduced to turning a crank it is no longer amusing nor truly profitable; but let this business become very profitable in a pecuniary sense, and so be "driven," as the phrase is, and carried on on a large scale, and the man is sunk in it, while only the pail or tray floats; we are interested in it only in the same way as the proprietor or company is.

OCTOBER 19, 1858

They may make equally good pails, and cheaper as well as faster, at the pail-factory with the home-made ones, but that interests me less, because the man is turned partly into a machine there himself. In this case, the workman's relation to his work is more poetic, he also shows more dexterity and is more of a man. You come away from the great factory saddened, as if the chief end of man were to make pails; but, in the case of the countryman who makes a few by hand, rainy days, the relative importance of human life and of pails is preserved, and you come away thinking of the simple and helpful life of the man . . .

OCTOBER 19, 1858

So completely emasculated and demoralized is our river that it is even made to observe the Christian Sabbath, and Hosmer tells me that at this season on a Sunday morning (for then the river runs lowest, owing to the factory and mill gates being shut above) little gravelly islands begin to peep out in the channel below. Not only the operatives make the Sunday a day of rest, but the river too, to some extent, so that the very fishes feel the influence (or want of *influence*) of man's religion . . . All nature begins to work with new impetuosity on Monday.

JULY 20, 1859

The young men of Concord and in other towns do not walk in the woods, but congregate in shops and offices . . . Their strongest attraction is toward the mill-dam [a local gathering place for news] . . . It is hard for the young, aye, and the old, man in the outskirts to keep away from the mill-dam a whole day; but he will find some excuse, as an ounce of cloves that might be wanted, or a *New England Farmer* [newspaper] still in the office, to tackle up the horse, or even go afoot, but he will go at some rate.

FEBRUARY 1851

The old [abandoned] house on Conantum is fast falling down.

OCTOBER 26, 1855

I sometimes see well-preserved [stone] walls running straight through the midst of high and old woods, built, of course, when the soil was cultivated many years ago, and am surprised to see slight stones still lying one upon another, as the builder placed them, while this huge oak has grown up from a chance acorn in the soil.

NOVEMBER 9, 1850

Now only a dent in the earth marks the site of most of these human dwellings; sometimes the well-dent where a spring oozed, now dry and tearless grass, or covered deep,—not to be discovered till late days by accident,—with a flat stone under the sod. These dents, like deserted fox-burrows, old holes, where once was the stir and bustle of human life overhead, and man's destiny, "fate, free-will, foreknowledge absolute," were all by turns discussed.

Still grows the vivacious lilac for a generation after the last vestige else is gone, unfolding still its early sweet-scented blossoms in the spring, to be plucked only by the musing traveller; planted, tended, weeded [?], watered by children's hands in front-yard plot,—now by wall-side in retired pasture, or giving place to a new rising forest. The last of that stirp, sole survivor of that family. Little did the dark children think that that weak slip with its two eyes which they watered would root itself so, and outlive them, and house in the rear that shaded it, and grown man's garden and field, and tell their story to the retired wanderer a half-century after they were no more,— blossoming as fair, smelling as sweet, as in that first spring. Its still cheerful, tender, civil lilac colors.

The woodland road, though once more dark and shut in by the forest, resounded with the laugh and gossip of inhabitants, and was notched and dotted here and there with their little dwellings. Though now but a humble rapid passage to neighboring villages or for the woodman's team, it once delayed the traveller longer, and was a lesser village in itself.

C. 1845–1847

The era of wild apples will soon be over. I wander through old orchards of great extent, now all gone to decay, all of native fruit which for the most part went to the cider-mill. But since the temperance reform and the general introduction of grafted fruit, no wild apples, such as I see everywhere in deserted pastures, and where the woods have grown up among them, are set out. I fear that he who walks over these hills a century hence will not know the pleasure of knocking off wild apples.

NOVEMBER 16, 1850

Old Mr. Joseph Hosmer, who lives where Hadley did, remembers when there were two or three times as many inhabitants in that part of the town as there are now: a blacksmith with his shop in front where he now lives, a goldsmith (Oliver Wheeler?) at the fork in the road just beyond him, one *in front* of Tarbell's, one in the orchard on the south side of the lane in front of Tarbell's, one, Nathan Wheeler, further on the right of the old road by the balm-of-Gilead, three between Tarbell's and J. P. Brown's, a tavern at Loring's, a store at the Dodge cottage that was burnt, also at Derby's (?), etc., etc. The farms were smaller then. One man now often holds two or three old farms. We walk in a deserted country.

NOVEMBER 21, 1851

How many are now standing on the European coast whom another spring will find located on the Red River, or Wisconsin! To-day we live an antediluvian life on our quiet homesteads, and to-morrow are transported to the turmoil and bustle of a crusading era.

Think how finite after all the known world is. Money coined at

Philadelphia is a legal tender over how much of it! You may carry ship biscuit, beef, and pork quite round to the place you set out from. England sends her felons to the other side [Australia] for safe keeping and convenience.

MARCH 27, 1840

I wrote a letter for an Irishman night before last, sending for his wife in Ireland to come to this country. One sentence which he dictated was, "Don't mind the rocking of the vessel, but take care of the children that they be not lost overboard."

MARCH 8, 1854

The Succession of Forest Trees

Observing the young pitch pines by the road south of Loring's lot that was so heavily wooded, George Hubbard remarked that if they were cut down oaks would spring up, and sure enough, looking across the road to where Loring's white pines recently stood so densely, the ground was all covered with young oaks.

APRIL 28, 1856

When Thoreau is acknowledged at all by scientists in the fields of ecology and forestry, it is inevitably in reference to his writings on the topic of succession. First presented to a group of farmers at a meeting of the Middlesex Agricultural Society, Thoreau's talk "The Succession of Forest Trees" was based on his extensive notes and journal entries; these in turn were based on an almost obsessive series of field observations of the changes that occurred on abandoned fields as they developed into forests and the subsequent changes in the forests as they were cut. For ecologists, succession, or the progressive recovery and change in vegetation following natural or human disturbance, has been a central topic of interest since the late 1800s, and in America it became the focus of one of the classic intellectual debates in the discipline. Disturbance is pervasive in the natural and cultural environment, and thus understanding the dynamics of forest ecosystems is a critical and fundamental issue for scientists, land managers, and naturalists. Thoreau's coining of the term "succession," his extensive series of observations on forest dynamics, and his interest in changes in nature have led a few ecologists, and many historical and literary scholars, to link him to the field of ecology.

Thoreau's studies of succession do have certain qualities that bind them historically to early ecologists' concept of the process, and both focused on the changes that were occurring in the eastern United States as a result of the widespread abandonment of old fields and the natural invasion by native plants. Thoreau's focus was somewhat more narrow than that of his ecological successors, however; he concentrated primarily on two aspects of the process of forest change, notably the tendency for one kind of tree—pine—to be the pioneer species that established initially in a field as it was abandoned and then for a second species—oak—to "succeed" the pine when it was cut or blown down. In contrast to later ecologists, Thoreau fit few other plants, such as mosses, ferns, herbs, or shrubs, into this scheme, and he paid scant attention to a process that is much more common today than it was in his landscape: the long-term natural development of forests in the absence of cutting.

Although he paid little attention to succession in his early walks, Thoreau's journal overflows with surprisingly repetitious descriptions of pine and oak successions by the early 1860s. Whether this shift reflects a change in Thoreau toward more detailed scientific inquiry is unclear, although this thesis has provided the focus for extended literary and biographical discussion. In any event, it is abundantly clear that by 1860 the process of old-field succession was widespread across the New England landscape as a consequence of broad-scale farm abandonment. It is also true that few of these "new" forests were allowed to develop naturally for long. Once they reached an age of 20 to 40 years, they were cut for a variety of wood products.

Thoreau's writing illuminates the important first stages of land abandonment. The process was not abrupt and wholesale, but gradual, inadvertent, and pulselike. Tilled field went to pasture, where it supported scraggly livestock as the pasture then went to shrubland and finally to forest. A struggle often ensued. Cows would continue to graze the grasses as the pines invaded the pastures. The farmer, though committing less and less effort to farming, would periodically beat back the pines with a bushwhack, scythe, or axe in order to maintain some grazing land for his dwindling herd. But eventually the new trees would overtop and crowd out the cows to win the battle. The farmer, sensing the futility and inevitability of the process, would drive the cattle out and would reluctantly accept his role as owner of a new woodlot.

From some of the details of Thoreau's writings we learn much about the processes that shaped the forests as they invaded the abandoned fields and came to dominate our landscape. For example, Thoreau's observation that the farmer's custom was to abandon land gradually and to let pines and cows coexist in pastures is significant. Although pines have many natural features that enable them to establish and grow in open grassy areas, such as small, easily dispersed seeds, tolerance for drought under full sunlight, and the ability to compete with dense grass sod, one additional advantage that they hold over hardwoods such as birch and maple is that both white and pitch pine are less palatable to grazing animals. Thus the practice of allowing cattle to continue grazing fields as they went into disuse and became forests may have been an important factor contributing to the prevalence of pitch pine and white pine on old fields, the development of the New England landscape into a region of pine forest, and the subsequent era of pine logging. As Thoreau correctly stated, social change and land abandonment resulted in a peculiarly New England mode of developing a woodlot.

Thoreau's observations and interpretations of the natural processes that occurred in the old fields of Concord are paralleled by observations made today and thus shed light on the development of our modern landscape. The first tree species to establish on an open site, such as a pasture that was no longer mowed or a harvested field that had been abandoned, Thoreau termed "pioneers" (a term still used by ecologists with very similar meaning). The particular species of pioneer that established on a site, he noted, depended on the soil conditions, the history of the site, and the abundance and dispersal ability of trees in the adjacent woods or fencerows. In pastures, Thoreau noted that white pine, pitch pine, and red cedar were the most common pioneers, with seedlings of apple trees occasionally bursting forth from the manure piles left by cows that had wandered previously through an adjacent orchard. If the site had been plowed recently, red maple or birch often dominated. However, once a young forest or shrubland developed, then oak generally appeared in the understory. Thoreau explained that these dynamics were related to the mechanisms of seed dispersal: wind in the case of the pines and maple, crows for juniper, cows for apple, and squirrels for oak. For the common species, Thoreau resorted to a military analogy to explain the rate and progress of movement of the different forests across the New England landscape: pines were the light infantry that took the longest strides and used wind to run ahead of the field of battle, advancing into open fields or hiding among rocks and along fences. Meanwhile, oaks were the grenadiers that marched deliberately behind, their acorns carried by the squirrels and jays who hid within

the cover of a developing pine woodland, incessantly advancing while bringing up the rear.

These insights enabled Thoreau to identify the origins and history of many forests upon first viewing. Pine woods or those composed of red cedar, birch, and maple tended to be "new forests" consisting of the first generation of trees on a previously cleared site. Many of these forests had abrupt, linear boundaries with adjoining stands of different structure and composition that marked property lines or land-use divisions. Many also contained telltale signs of their agricultural past, including pasture sod and abundant grasses as well as weedy species such as blackberries and lady's slipper that established rapidly on abandoned sites. In contrast, oaks generally represented a second-generation forest since they often seeded into established pine woods and germinated in the carpet of needles beneath the pines. When the pines were removed or died, the oaks emerged as the next forest.

Thoreau dwells intently on the processes leading to this second generation of oak. He recounts in detail the tendency for oak to establish beneath pines, which he attributes in part to their reliance for dispersal on animals such as squirrels and chipmunks that need the pines for cover, and in part to the ability of oak to persist where light and air are in demand. Acorns, once transported and cached in the woods, may be forgotten by the animals and emerge as the small seedlings that are common in pine forests. Meanwhile, the pines (especially pitch pine) are much less tolerant of shade than oak or many other hardwoods, and consequently their seedlings do not establish in large numbers within a forest. The oak seedlings easily persist for eight to ten years beneath pine, and therefore if the stand is cut (the typical fate in the mid-1800s), is burned, or is blown down by very strong winds, the next forest would be oak. The natural process is slow, perhaps requiring centuries because of the longevity of the pines, but in his landscape interpretations Thoreau was always willing to invoke long ecological and geological time scales. He recognized that the abundant seed production and ongoing seed collection and caching by generations of chipmunks or squirrels could easily account for all the oak seedlings that he located in a pine forest.

Thus, through his descriptions of succession Thoreau identified a prevalent phenomenon that helped to explain the increasing abundance of pine forest in his lifetime and its gradual decline in ours. In fact, the extent of old-field white pine forest became so great across the northeastern United States that forest maps constructed at the turn of the century by Charles Sprague Sargent at Harvard University and George Nichols at Yale University identified much of New England as the "White Pine Region." By the mid-twentieth century, however, the conversion of these pine areas to oak and other hardwood forests through the process of cutting and succession identified by Thoreau was convincing evidence that extensive white pine forests were a transient phenomenon generated by cultural and natural history. This has led ecologists and foresters to remap the region in the 1950s as natural hardwood forest with a mixture of hemlock and pine.

Although less common today, the process of forest succession continues in thousands of fields and woodlands across New England, and we can observe its consequences in the landscape patterns it has created. From an airplane or when looking across a broad river valley in areas still being farmed, we can frequently view a forest of white pine situated between fields on one side and a hardwood forest of oak or maple on the other. This first generation of pine has filled a pasture most recently abandoned by the farmer or his predecessor, and a careful search will generally show a second generation of hardwood seedlings and saplings developing beneath. Meanwhile, on the ground below the adjoining oak forest, the rotting stumps of old pines confirm Thoreau's observations and document that this modern hardwood forest was previously pine, and before that an open field. Thus, the historical sequence of field, abandoned field, pine wood, and hardwood forest is displayed as a spatial pattern in the modern countryside.

In managing forest areas today, we can use Thoreau's knowledge of succession and landscape history to shape our expectations for their future condition or to modify the approach we take. For many stands of pine and other pioneers, we can anticipate that with time and natural development the vegetation will become something very different. The seedlings and saplings below the pine offer a major clue as to what the stand will become. In fact, if the generations of foresters and ecologists who followed Thoreau had actually read and studied his work, they would have saved a great deal of effort. Both groups spent the first few decades of the twentieth century rediscovering the successional processes that Thoreau describes so well, and they could have benefited greatly from the empirical observations he offered as well as his insistence that the process could vary as a consequence of human history, plant biology, natural environment, and the dispersal capabilities of the local plants. In particular, the foresters who spent the first half of this century trying to get white pine to establish after it was cut could have been spared years of futile effort had they studied Thoreau's observations. The particular forest changes that have occurred in the New England landscape over the last century were actually a historical accident that can be best understood through a strong knowledge of plant biology and an awareness of cultural history.

OLD FIELDS AND PIONEER SPECIES

What shall we say to that management that halts between two courses,—does neither this nor that, but botches both? I see many a pasture on which the pitch or white pines are spreading, where the bush-whack is from time to time used with a show of vigor, and I despair of my trees,—I say mine, for the farmer evidently does not mean they shall be his,—and yet this questionable work is so poorly done that those very fields grow steadily greener and more forest-like from year to year in spite of cows and bush-whack, till at length the farmer gives up the contest from sheer weariness, and finds himself the owner of a wood-lot. Now whether wood-lots or pastures are most profitable for him I will not undertake to say, but I am certain that a wood-lot and pasture combined is not profitable.

OCTOBER 18, 1860

I see, on the side of Fair Haven Hill, pines which have spread, apparently from the north, one hundred rods, and the hillside begins to wear the appearance of woodland, though there are many cows feeding amid the pines. The custom with us is to let the pines spread thus into the pasture, and at the same time to let the cattle wander there and contend with the former for the possession of the ground, from time to time coming to the aid of the cattle with a bush-whack. But when, after some fifteen or twenty years, the pines have fairly prevailed over us both, though they have suffered terribly and the ground is strewn with their dead, we then suddenly turn about, coming to the aid of the pines with a whip, and drive the cattle out. They shall no longer be allowed to scratch their heads on them, and we fence them in. This is the actual history of a great many of our wood-lots. While the English have taken great pains to learn how to create forests, this is peculiarly our mode. It is plain that we have thus both poor pastures and poor forests.

OCTOBER 19, 1860

The pastures on this hill and its spurs are sprinkled profusely with thorny pyramidal apple scrubs, very thick and stubborn, first planted by the cows, then browsed by them and kept down stubborn and thorny for years, till, as they spread, their centre is protected and beyond reach and shoots up into a tree, giving a wine-glass form to the whole; and finally perchance the bottom disappears and cows come in to stand in the shade and rub against and redden the trunk. They must make fine dark shadows, these shrubs, when the sun is low; perfectly pyramidal they are now, many of them. You see the cow-dung everywhere now with a hundred little trees springing up in it. Thus the cows create their own shade and food.

MAY 22, 1853

I see in many places little barberry bushes just come up densely in the cow-dung, like young apple trees, the berries having been eaten by the cows. Here they find manure and an open space for the first year at least, when they are not choked by grass or weeds. In this way, evidently, many of these clumps of barberries are commenced.

JUNE 28, 1858

How suddenly, after all, pines seem to shoot up and fill the pastures! I wonder that the farmers do not earlier encourage their growth. To-day, perchance, as I go through some run-out pasture, I observe many young white pines dotting the field, where last year I had noticed only blackberry vines; but I see that many are already destroyed or injured by the cows which have dived into them to scratch their heads or for sport (such is their habit; they break off the leading shoot and bend down the others of different evergreens), or perchance where the farmer has been mowing them down, and I think the owner would rather have a pasture here than a wood-lot. A year or two later, as I pass through the same field, I am surprised to find myself in a flourishing young wood-lot, from which the cows are now carefully fenced out, though there are many open spaces, and I perceive how much further advanced it would have been if the farmer had been more provident and had begun to abet nature a few years earlier. It is surprising by what leaps—two or three feet in a season—the pines stretch toward the sky, affording shelter also to various hardwoods which plant themselves in their midst.

MAY 19, 1857

Two or three pines will run swiftly forward a quarter of a mile into a plain, which is their favorite field of battle, taking advantage of the least shelter, as a rock, or fence, that may be there, and intrench themselves behind it, and if you look sharp, you may see their plumes waving there. Or, as I have said, they will cross a broad river without a bridge, and as swiftly climb and permanently occupy a steep hill beyond.

OCTOBER 19, 1860

How commonly you see pitch pines, white pines, and birches filling up a pasture, and, when they are a dozen or fifteen years old, shrub and other oaks beginning to show themselves, inclosing apple trees and walls and fences gradually and so changing the whole aspect of the region . . . The cow-paths, the hollows where I slid in the winter, the rocks, are fast being enveloped and becoming rabbit-walks and hollows and rocks in the woods.

NOVEMBER 25, 1860

There is the pitch pine field northeast of Beck Stow's Swamp, where some years ago I went a-blackberrying and observed that the pitch pines were beginning to come in, and I have frequently noticed since how fairly they grew, dotting the plain as evenly as if dispersed by art. To-day I was aware that I walked in a pitch pine wood, which ere long, perchance, I may survey and lot off for a wood auction and see the choppers at their work. There is also the old pigeon-place field by the Deep Cut. I remember it as an open grassy field. It is now one of our most pleasant woodland paths.

NOVEMBER 9, 1850

Came through the pine plains behind James Baker's, where late was open pasture, now open pitch pine woods, only here and there the grass has given place to a carpet of pine-needles. These are among our pleasantest woods,—open, level, with blackberry vines interspersed and flowers, as lady's-slippers, earlier, and pinks on the outskirts.

JULY 16, 1851

There are no more beautiful natural parks than these pastures in which the white pines have sprung up spontaneously, standing at handsome intervals, where the wind chanced to let the seed lie at last, and the grass and blackberry vines have not yet been killed by them.

JANUARY 30, 1860

Bathing on the side of the deep cove [at Walden Pond], I noticed just below the high-water line (of rubbish) quite a number of little pines which have just sprung up amid the stones and sand and wreck, some with the seed atop. This, then, is the state of their coming up naturally.

JULY 20, 1860

At my Swamp Brook crossing at Willow Bay, I see where a great many little red maples have sprung up in a potato-field, apparently since the last plowing or cultivating this year. They extend more or less thickly as much as eleven rods [one rod equals 16 1/2 feet] in a northwest direction from a small tree, the only red maple in that neighborhood. And it is evidently owing to the land having been cultivated this year that the seed vegetated there; otherwise there would now be no evidence that any such seeds had fallen here. Last year and for many years it has been a pasture. It is evident that land may be kept as a pasture and covered with grass any number of years, and though there are maples adjacent to it, none of the seed will catch in it; but at last it is plowed, and this year the seed which falls on it germinates, and if it chances not to be plowed again, and cattle are kept out, you soon have a maple wood there. So of other light-seeded trees.

SEPTEMBER 4, 1860

I notice that the first shrubs and trees to spring up in the sand on railroad cuts in the woods are sweet-fern, birches, willows, and aspens, and pines, white and pitch; but all but the last two chiefly disappear in the thick wood that follows. The former are the pioneers.

OCTOBER 22, 1860

SUCCESSION TO OAK FOREST

I see many seedling shrub oaks springing up in Potter's field by the swamp-side, some (of last year) in the open pasture, but many more in the birch wood half a dozen rods west from the shrub oaks by the path. The former were dropped by the way. They plant in birch woods as in pines. This small birch wood has been a retreat for squirrels and birds. When I examine the little oaks in the *open* land there is always an effete acorn with them.

SEPTEMBER 7, 1860

Apparently the pine woods are a natural nursery of oaks, from whence we might easily transplant them to our grounds, and thus save some of those which annually decay, while we let the pines stand. Experience has proved, at any rate, that these oaks will bear exposure to the light. It is remarkable that for the most part there are no seedling oaks in the open grassy fields and pastures.

I examine under the pitch pines by Thrush Alley to see how long the oaks live under dense pines. The oldest oaks there are about eight or ten years old . . . This corresponds exactly with the experience of the English planters, who begin to shred [thin] the branches of the nursing pines when the oaks are six or seven years old and to remove the pines altogether when the oaks are eight to ten years old.

But in openings amid the pines, though only a rod in diameter, or where the pines are thin, and also on their edges, the oaks shoot up higher and become trees, and this shows how mixed woods of pine and oak are produced . . . Or when you thin out pine wood, the oaks spring up here and there . . .

It is surprising how many accidents these seedling oaks will survive. We have seen [?] that they commonly survive six to ten years under the thickest pines and acquire stout and succulent roots. Not only [do] they bear the sudden exposure to the light when the pines are cut, but, in case of a more natural succession, when a fire runs over the lot and kills pines and birches and maples, and oaks twenty feet high, these little oaks are scarcely injured at all, and they will still be just as high the next year, if not in the fall of the same year if the fire happens early in the spring. Or if in the natural course of events a fire does not occur nor a hurricane, the soil may at last be exhausted for pines, but there are always the oaks ready to take advantage of the least feebleness and yielding of the pines.

OCTOBER 17, 1860

SEEDS AND SEED DISPERSAL

I once found a kernel of corn in the middle of a deep wood by Walden, tucked in behind a lichen on a pine, about as high as my head, either by a crow or a squirrel. It was a mile at least from any corn-field.

NOVEMBER 19, 1850

Then there are the countless downy seeds (thistle-like) of the golden-rods, so fine that we do not notice them in the air. They cover our clothes like dust. No wonder they spread over all fields and far into the woods.

OCTOBER 24, 1860

I have often wondered how red cedars could have sprung up in some pastures which I knew to be miles distant from the nearest fruit-bearing cedar, but it now occurs to me that these and barberries, etc., may be planted by the crows, and probably other birds.

FEBRUARY 4, 1856

Pines take the first and longest strides. Oaks march deliberately in the rear.

The pines are the light infantry, *voltigeurs,* supplying the scouts and skirmishers; the oaks are the grenadiers, heavy-paced and strong, that form the solid phalanx.

OCTOBER 19, 1860

Suppose one were to take such a boxful of birch seed as I have described into the meeting-house belfry in the fall, and let some of it drop in every wind, but always more in proportion as the wind was stronger, and yet so husband it that there should be some left for every gale even till far into spring; so that this seed might be blown toward every point of the compass and to various distances in each direction. Would not this represent a single birch tree on a hill? Of which trees (though only a part on hills) we have perhaps a million. And yet some feel compelled to suppose that the birch trees which spring up after a burning are spontaneously generated—for want of seed! It is true [it] does not come up in great quantities at the distance I have spoken of, but, if only one comes up there this year, you may have a million seeds matured there a few years hence.

It is true that the greater part of these seeds fall near the trees which bore them, and comparatively few germinate; yet, when the surface is in a favorable condition, they may spring up in very unexpected places.

MARCH 8, 1861

If the pine seed is spontaneously generated, why is it not so produced in the Old World as well as in America? I have no doubt that it can be raised from the seed in corresponding situations there, and that it will seem to spring up just as mysteriously there as it does here. Yet, if it will grow so *after* the seed has been carried thither, why should it not before, if the seed is unnecessary to its production?

MARCH 22, 1861

As a consequence of the different manner in which trees which have winged seeds and those which have not are planted,—the [former] being blown together in one direction by the wind, the latter being dispersed irregularly by animals,—I observe that the former, as pines (which (the white) are said in the primitive wood to grow in communities), white birches, red maples, alders, etc., often grow in more or less regular rounded or oval or conical patches, as the seeds fell, while oaks, chestnuts, hickories, etc., simply form woods of greater or less extent whether by themselves or mixed; *i. e.,* they do not naturally spring up in an oval form (or elliptical) unless they derive it from the pines under which they were planted.

OCTOBER 16, 1860

If any one presumes that, after all, there cannot be so many nuts planted as we see oaks spring up at once when the pines are cut, he must consider that *according to the above calculation* (two pages back) there are some ten years for the animals to plant the oak wood in; so that, if the tract is ten rods square or contains one hundred square rods, it would only be necessary that they should plant ten acorns in a year which should not be disturbed, in order that there might be one oak to every square rod at the end of ten years. This, or anything like this, does not imply any very great activity among the squirrels. A striped squirrel [i.e., a chipmunk] could carry enough in his cheeks at one trip.

OCTOBER 17, 1860

See how an acorn is planted by a squirrel, just under a loose covering of moist leaves where it is shaded and concealed, and lies on its side on the soil, ready to send down its radicle next year.

OCTOBER 18, 1860

This reminds me that, though I often see little white pines under pines and under oaks, I rarely if ever (unless I am mistaken) see many young pitch pines there. How is it? Do the pitch pines require more light and air?

OCTOBER 22, 1860

As I am coming out of this, looking for seedling oaks, I see a jay, which was screaming at me, fly to a white oak eight or ten rods from the wood in the pasture and directly alight on the ground, pick up an acorn, and fly back into the woods with it. This was one, perhaps the most effectual, way in which this wood was stocked with the numerous little oaks which I saw under that dense white pine grove. Where will you look for a jay sooner than in a dense pine thicket? It is there they commonly live and build.

What if the oaks are far off? Think how quickly a jay can come and go, and how many times in a day!

OCTOBER 27, 1860

So far as our noblest hardwood forests are concerned, the animals, especially squirrels and jays, are our greatest and almost only benefactors. It is to them that we owe this gift. It is not in vain that the squirrels live in or about every forest tree, or hollow log, and every wall and heap of stones.

OCTOBER 31, 1860

We need not be surprised at these results when we consider how persevering Nature is, and how much time she has to work in, though she works slowly. A great pine wood may drop many millions of seeds in one year, and if only half a dozen are conveyed a quarter of a mile and lodge against some fence, and only one comes up and lives there, yet in the course of fifteen or twenty years there are fifteen or twenty young trees there, and they begin to make a show and betray their origin. It does not imply any remarkable rapidity or success in Nature's operations.

NOVEMBER 8, 1860

A writer in the *Tribune* speaks of cherries as one of the trees which come up numerously when the forest is cut or burned, though [cherries were] not known there before. This may be true because there was no one knowing in these matters in that neighborhood. But I assert that it *was* there before, nevertheless; just as the little oaks are in the pine woods, but never grow up to trees till the pines are cleared off. Scarcely any plant is more sure to come up in a sprout-land here than the wild black cherry, and yet, though only a few inches high at the end of the first year after the cutting, it is commonly several years old, having maintained a feeble growth there so long. There is where the birds have dropped the stones [cherry seeds], and it is doubtful if those dropped in pastures and open land are as likely to germinate.

MARCH 22, 1861

At first, perchance, there would be an abundant crop of rank garden weeds and grasses in the cultivated land,—and rankest of all in the cellar-holes,—and of pinweed, hardhack, sumach, blackberry, thimble-berry, raspberry, etc., in the fields and pastures. Elm, ash, maples, etc., would grow vigorously along old garden limits and main streets. Garden weeds and grasses would soon disappear. Huckleberry and blueberry bushes, lambkill, hazel, sweet-fern, barberry, elder, also shad-bush, choke-berry, andromeda, and thorns, etc., would rapidly prevail in the deserted pastures. At the same time the wild cherries, birch, poplar, willows, checkerberry would reëstablish themselves. Finally the pines, hemlock, spruce, larch, shrub oak, oaks, chestnut, beech, and walnuts would occupy the site of Concord once more. The apple and perhaps all exotic trees and shrubs and a great part of the indigenous ones named above would have disappeared, and the laurel and yew would to some extent be an underwood here, and perchance the red man once more thread his way through the mossy, swamplike, primitive wood.

NOVEMBER 23, 1860

Losses and Change

Animals: From Bobolinks to Bears

As I proceed along the back road I hear the lark still singing in the meadow, and the bobolink, and the gold robin on the elms, and the swallows twittering about the barns. A small bird chasing a crow high in the air, who is going home at night. All nature is in an expectant attitude.

JUNE 14, 1851

But when I consider that the nobler animals have been exterminated here,—the cougar, panther, lynx, wolverene, wolf, bear, moose, deer, the beaver, the turkey, etc., etc.,—I cannot but feel as if I lived in a tamed, and, as it were, emasculated country. Would not the motions of those larger and wilder animals have been more significant still? Is it not a maimed and imperfect nature that I am conversant with? As if I were to study a tribe of Indians that had lost all its warriors.

MARCH 23, 1856

Conservation biologists in eastern North America are witnessing and trying to manage two opposing trends among native wildlife. While the numbers and types of forest-dwelling mammals, birds, amphibians, and insects are increasing at a phenomenal rate and reaching levels not seen since colonial days, the delightful birds and butterflies of the open fields and shrublands are disappearing. The regional proportions of forest and agriculture are reversed from those of Henry Thoreau's day, and the populations of wildlife are gradually reflecting this change. Thoreau's daily observations of animals in the cultural landscape of the nineteenth century can give us perspective and guidance in understanding the changes that are occurring around us today.

One of the most remarkable and yet relatively unheralded natural history phenomena in the twentieth century has been the reestablishment and proliferation of many forest-dwelling animal species across eastern North America. Deer, turkey, beaver, bear, coyote, pileated woodpeckers, and fisher are thriving or increasing. Moose, eagles, and bobcat are encountered fairly frequently, and the occasional claims of cougar sightings in central and northern New England and the Adirondacks are matched by the recent confirmation of a wolf in Maine. The rate of increase in many large mammals over recent decades has transformed what was previously rare and notable into the widespread and common. While this resurgence of native animals has given a wilder, more natural feeling to the New England landscape, it has also generated problems for a growing suburban population that is relatively dissociated from the land and unaccustomed to dealing with the consequences of encounters with wildlife.

Beaver, which had been eliminated from New England by extensive trapping during early colonial times, have expanded widely across the region in the last three decades. In many areas beaver are actually much more abundant than at any time since European settlement. In the absence of major predators and trapping, they may even exceed previous densities in a few habitats. This reappearance of beavers along eastern streams has restored many natural dams, meadows, and ponds that have been missing for more than three hundred years from these watery ecosystems. They also accentuate the landscape-level diversity of natural ecosystems and provide important habitats for waterfowl, amphibians, insects, and many additional mammals. At the same time, the increase in beavers along rivers, tributaries, and water bodies of all sizes together with their natural inclination to fell trees, raise the existing water table, flood surrounding upland areas, and create wetlands poses complex problems for landowners, municipal planners, and natural resource agencies. As beavers increase in numbers and range we are faced with the dilemma of either accepting their constant re-engineering of the landscape or increasing our efforts to control them.

Canada geese, whose elegant V's and distant honking are memorable during the crisp, clear days of spring and fall, were uncommon fifty years ago. Recently, however, they have expanded their numbers and habitat tremendously and now reside year-round on many golf courses, lawns of corporate headquarters, school ball fields, municipal parks, and reservoirs. Here they create constant challenges for caretakers, municipal workers, and water quality professionals, who try to drive them and the similarly resurgent seagulls away.

The burgeoning deer population across southern New England greatly enhances the atmosphere of our reforesting landscape and offers many people their first exposure to large wildlife. However, large numbers of deer also create a dilemma for suburban dwellers, who are generally opposed to hunting in their backyards and yet are eager to protect their gardens and ornamental plants and to maintain productive, healthy forests. Similarly, the southern expansion of a growing moose population has prompted the establishment of a new hunting season in Vermont and programs for driver education across central and even southern New England (for example, "Brake for Moose" bumper stickers and newly erected road signs in Massachusetts), while generating increasing problems for animal control officers along major interstate highways and in the Boston suburbs.

These recent phenomena contrast sharply with the status of wildlife populations in the nineteenth-century landscape described by Thoreau. Remarkably, for Thoreau and his acquaintances the muskrat was the largest mammal to be seen commonly in eastern Massachusetts, and deer were so rare that a sighting of one in the town of Tewksbury warranted special mention in his journal. The combination of deforestation, habitat destruction, and unbridled hunting and trapping of animals had led to widespread decimation and local extinction by the early 1800s. To Thoreau, this loss of major animal species (the "noble" animals—moose, wolf, deer, beaver, fisher, among others) produced a strong sense of impoverishment and artificiality in his landscape. The elimination of these animals as a result of human activity also provided him with one of the strongest examples of the debasement and short-sightedness of mankind.

Thoreau attributed the change in wildlife populations to two broad causes—human alteration of the natural environment and rapacious slaughter. The conversion of forest to field, the repeated cutting of woodlots, the damming of major riverways, and the ditching and draining of wetlands all led to the loss of habitat and animal populations. Thoreau railed against such "improvements" wrought by humans, for they inevitably harmed native animal species and despoiled natural habitats. No example of this change was more striking to him than the alteration of rivers and the associated decline in migrating fish—the salmon, shad, herring, and alewife. These particular fish are anadromous species that require free

movement from the upper reaches of the streams where they hatch, to the oceans where they live for one or more years, and then back to the stream headwaters where they spawn. The massive dams and factories that appeared during Thoreau's day represented the increasing human development activity that peaked with the Industrial Revolution. By restricting the movement of migrating fish and clogging the waters with clouds of manufacturing and human waste, the industries and growing cities poisoned the environment, produced thick blooms of algae that choked the water in response to increases in nutrients, and deprived the streams of a vital living quality. Thoreau felt that the slow-moving Concord River, which was dammed downstream in the Merrimack River at the industrial cities of Lowell and Lawrence, could be considered literally dead.

Thoreau's descriptions of the direct impact of humans on animal populations provide an interesting combination of dispassionate reporting on country life and vitriolic condemnation of human morals. The ubiquity of hunting and trapping among Thoreau's acquaintances and the importance of these enterprises to nineteenth-century life allowed Thoreau to view and report on their daily successes as commonplace occurrences. Indeed, in his role as neighborhood natural historian, Thoreau was often the recipient of the inedible, though frequently fascinating, trophies that resulted from the capricious hunting activities of his Concord acquaintances. Thus, as he marveled in his writings over the paws, pelts, talons, and carcasses that were regularly delivered to his doorstep, Thoreau often appeared to accept the killing of woodpeckers, hawks, turtles, and even lynx with little regard. It was an inescapable fact of the country life of New England that many animals were killed routinely, that poisoning of perceived vermin species was common, that squirrel skins, crow wings, and muskrat tails were tacked by the dozens to barn walls as minor trophies, that turtle meat was eaten alongside beef and vegetables, that species in decline such as the passenger pigeon were still baited and trapped.

In his occasional reflections, however, Thoreau railed strongly against the human onslaught on wildlife because it seemed so unnecessarily driven by greed and hypocrisy and led to such an impoverished state of nature. The wholesale elimination of large native mammals, driven by such petty and ridiculous fads as the one for skunk fur that emerged from some nameless salon of couture in the spring of 1859, was singled out by Thoreau as an example of the woeful condition of humankind. He believed that European man had initially debased the Indians by encouraging trapping and hunting and then, through his own pursuits in the New World, had stooped to an even lower level, which Thoreau equated to that of a bone and rag picker. The hypocrisy that allowed "civilized man" to eliminate other species was described strikingly by Thoreau in the actions of the self-proclaimed progressive individual who nonetheless insisted on wearing animal skins, and in the political system that established wildlife laws based on the utility of a species or offered bounties based on the perceived negative impacts of the animal on human enterprise.

In sharp contrast to his pessimistic review of nature's loss, Thoreau's delightful descriptions of the diverse bird life of his time present us with views of wildlife scenes that are quite unknown in our modern landscape. Whereas most of the larger and wilder animal species were absent from nineteenth-century New England, many of the species of birds that were very familiar to Thoreau are considerably less abundant today. Most notable in his journal entries is the prominence of meadow, field, and shrubland bird species such as the bobolink, bluebird, meadowlark, field sparrow, and song sparrow. He identifies the song sparrow as the most familiar New England bird and a species commonly heard in fields and pastures. Bobolinks, larks, and bluebirds are described in dense flocks, covering fence rails and alder thickets. Many of these and other species that Thoreau describes in his daily journals have declined dramatically in the northeastern United States over the last century and have emerged as major priorities for conservation by regional and national environmental groups. Species of specific concern include the bittern, whip-poor-will, eastern bluebird, bobolink, and eastern meadowlark. Each of these thrived in the pastoral landscape of Thoreau, where swallows twittered incessantly around barns and the American bittern ("stake-drivers") boomed daily at dusk across the meadows.

Thus, as the landscape that provides the habitat, shelter, and food for animal species has changed from Thoreau's day to the present, the wildlife has changed accordingly. Thoreau clearly recognized that animal species are finely attuned to the structure and composition of vegetation. As farm abandonment progressed, the pastures, grain fields, and meadows of Thoreau's day developed into fields of weeds, then thickets, and finally forests, and the bobolink and quail were replaced by the turkey and spruce grouse. As trees became larger, New England saw the return of the pileated woodpecker, a crow-sized bird that carves deep cavities for food and nests in larger trees, and the great blue heron, which forms rookeries composed of dozens of nests in the stark and leafless dead trees encircling beaver swamps that have flooded the forests. In today's landscape, where forest has almost completely replaced field and shrubland, the birds (and the butterflies, amphibians, and other animals) that Thoreau described are uncommon and rare indeed. If we lament this loss and seek to reverse this historical trend, we will need to maintain and expand the cultural elements of Thoreau's landscape. On the other hand, if we choose not to intervene, we can expect a continued loss of open field and shrubland species and an ongoing expansion of woodland wildlife.

Thoreau's passages on wildlife reveal a paradox of loss and gain. The sounds and sights of wildlife that brought him joy and marked the seasonal progression of his New England countryside were largely produced by species that thrived on the same landscape changes that proved inimical to the wilder species that he longed for. On the other hand, his tamed landscape offered food, shelter, and habitat to a diverse array of animals that we are gradually losing today as the fields and blackberry thickets and scrubby pastures of thistle, juniper, and goldenrod disappear. In their place have arrived numbers of other species, some original inhabitants of New England such as the bear and the moose, and others, such as the coyote, that are newcomers. They and the forests that harbor them remind us of the wilder side of our nature, a nature that Thoreau could only long for.

The lark sings in the meadow; the very essence of the afternoon is in his strain. This is a New England sound . . .

JULY 16, 1851

The song sparrow, the most familiar and New England bird, is heard in fields and pastures, setting this midsummer day to music, as if it were the music of a mossy rail or fence post; a little stream of song, cooling, rippling through the noon,—the usually unseen songster usually unheard like the cricket, it is so common,—like the poet's song, unheard by most men, whose ears are stopped with business, though perchance it sang on the fence before the farmer's house this morning for an hour.

JULY 16, 1851

The whip-poor-wills now begin to sing in earnest about half an hour before sunrise, as if making haste to improve the short time that is left them. As far as my observation goes, they sing for several hours in the early part of the night, are silent commonly at midnight,—though you may meet [them] then sitting on a rock or flitting silently about,—then sing again just before sunrise.

SEPTEMBER 9, 1851

From Conant's summit I hear as many as fifteen whip-poor-wills—or whip-or-I-wills—at once, the succeeding cluck sounding strangely foreign, like a hewer [woodworker] at work elsewhere.

JUNE 14, 1851

The bobolink sings descending to the meadow as I go along the railroad to the pond. The seringo-bird and the common song sparrow,—and the swallows twitter. The plaintive strain of the lark, coming up from the meadow, is perfectly adapted to the hour. When I get nearer the wood, the veery is heard, and the oven-bird, or whet-saw, sounds hollowly from within the recesses of the wood.

JUNE 28, 1852

I see a dense, compact flock of bobolinks going off in the air over a field. They cover the rails and alders, and go rustling off with a brassy, tinkling note like a ripe crop as I approach, revealing their yellow breasts and bellies. This is an autumnal sight, that small flock of grown birds in the afternoon sky.

AUGUST 15, 1852

I no sooner step out of the house than I hear the bluebirds in the air, and far and near, everywhere except in the woods, throughout the town you may hear them,—the blue curls of their warblings,—harbingers of serene and warm weather, little azure rills of melody trickling here and there from out the air, their short warble trilled in the air reminding of so many corkscrews assaulting and thawing the torpid mass of winter, assisting the ice and snow to melt and the streams to flow.

MARCH 18, 1853

How indispensable our one or two flocks of geese in spring and autumn! What would be a spring in which that sound was not heard? Coming to unlock the fetters of northern rivers. Those annual steamers of the air.

APRIL 15, 1852

Scared up three blue herons in the little pond close by, quite near us. It was a grand sight to see them rise, so slow and stately, so long and limber, with an undulating motion from head to foot, undulating also their large wings, undulating in two directions, and looking warily about them. With this graceful, limber, undulating motion they arose, as if so they got under way, their two legs trailing parallel far behind like an earthy residuum to be left behind. They are large, like birds of Syrian lands, and seemed to oppress the earth, and hush the hillside to silence, as they winged their way over it, looking back toward us. It would affect our thoughts, deepen and perchance darken our reflections, if such huge birds flew in numbers in our sky. Have the effect of magnetic passes. They are few and rare.

APRIL 19, 1852

After the warm weather has come, both morning and evening you hear the bittern pumping in the fens . . . Methinks that in the resemblance of this note to rural sounds, to sounds made by farmers, the protection, the security, of the bird is designed. Minott says: "I call them belcher-squelchers. They go *slug-toot, slug-toot, slug-toot.*"

JUNE 20, 1852

Heard a stake-driver [bittern] yesterday in the rain. It sounded exactly like a man pumping, while another man struck on the head of the pump with an axe, the last strokes sounding peculiarly dry and hard like a forcible echo from the wood-side. One would think all Concord would be built on piles by this time.

MAY 27, 1853

When setting the pines at Walden the last three days, I was sung to by the field sparrow. For music I heard their jingle from time to time. That the music the pines were set to, and I have no doubt they will build many a nest under their shelter. It would seem as if such a field as this—a dry open or half-open pasture in the woods, with small pines scattered in it—was well-nigh, if not quite, abandoned to this one alone among the sparrows. The surface of the earth is portioned out among them. By a beautiful law of distribution, one creature does not too much interfere with another. I do not hear the song sparrow here. As the pines gradually increase, and a wood-lot is formed, these birds will withdraw to new pastures, and the thrushes, etc., will take their place.

APRIL 22, 1859

Is not this [the muskrat] the heaviest animal found wild in this township?

MAY 17, 1854

Minott says his mother told him she had seen a deer come down the hill behind her house, where I. Moore's now is, and cross the road and the meadow in front; thinks it may have been eighty years ago.

JANUARY 21, 1853

Farmer says that a farmer in Tewksbury told him two or three years ago that he had seen deer lately on the pine plain thereabouts.

SEPTEMBER 13, 1860

Minott says that old Sam Nutting, the hunter,—Fox Nutting, Old Fox, he was called,—who died more than forty years ago (he lived in Jacob Baker's house, Lincoln; came from Weston) and was some seventy years old then, told him that he had killed not only bear about Fair Haven among the walnuts, but *moose!*

MARCH 10, 1853

Fishermen say that no fish can get above the dam at Lawrence. No shad, etc., were caught at Lowell last year.

APRIL 13, 1853

If salmon, shad, and alewives were pressing up our river now, as formerly they were, a good part of the villagers would thus, no doubt, be drawn to the brink at this season. Many inhabitants of the neighborhood of the ponds in Lakeville, Freetown, Fairhaven, etc., have petitioned the legislature for permission to connect Little Quitticus Pond with the Acushnet River by digging, so that the herring can come up into it. The very fishes in countless schools are driven out of a river by the *improvements* of the civilized man, as the pigeon and other fowls out of the air. I can hardly imagine a greater change than this produced by the influence of man in nature. Our Concord River is a dead stream in more senses than we had supposed. In what sense now does the spring ever come to the river, when the sun is not reflected from the scales of a single salmon, shad, or alewife? No doubt there is *some* compensation for this loss, but I do not at this moment see clearly what it is. That river which the aboriginal and indigenous fishes have not deserted is a more primitive and interesting river to me. It is as if some vital quality were to be lost out of a man's blood and it were to circulate more lifelessly through his veins. We are reduced to a few migrating (?) suckers, perchance.

APRIL 11, 1857

In the woods just this side [of the railroad], we came upon a partridge standing on the track, between the rails over which the cars had just passed. She had evidently been run down, but, though a few small feathers were scattered along for a dozen rods beyond her, and she looked a little ruffled, she was apparently more disturbed in mind than body. I took her up and carried her one side to a safer place. At first she made no resistance, but at length fluttered out of my hands and ran two or three feet. I had to take her up again and carry and drive her further off, and left her standing with head erect as at first, as if beside herself. She was not lame, and I suspect no wing was broken. I did not suspect that this swift wild bird was ever run down by the cars. We have an account in the newspapers of every cow and calf that is run over, but not of the various wild creatures who meet with that accident. It may be many generations before the partridges learn to give the cars a sufficiently wide berth.

APRIL 12, 1858

I hear from far the scream of a hawk circling over the Holden woods and swamp. This accounts for those two men with guns just entering it. What a dry, shrill, angry scream! I see the bird with my glass resting upon the topmost plume of a tall white pine. Its back, reflecting the light, looks white in patches; and now it circles again. It is a red-tailed hawk. The tips of its wings are curved upward as it sails. How it scolds at the men beneath! I see its open bill. It must have a nest there. Hark! there goes a gun, and down it tumbles from a rod or two above the wood.

APRIL 30, 1855

Went to Garfield's for the hawk of yesterday. It was nailed to the barn *in terrorem* and as a trophy. He gave it to me with an egg.

MAY 1, 1855

Returning on the crust, over Puffer's place, I saw a fine, plump hen hanging from an apple tree and a crow from another, probably poisoned to kill foxes with,—a hen which probably a fox had killed.

FEBRUARY 28, 1856

Coombs has a stand west of Nut Meadow, and he says that he has just shot fourteen hawks there, which were after the pigeons. I have one which he has shot within a day or two and calls a pigeon hawk.

SEPTEMBER 14, 1859

Yet what is the character of our gratitude to these squirrels, these planters of forests? We regard them as vermin, and annually shoot and destroy them in great numbers, because—if we have any excuse—they sometimes devour a little of our Indian corn, while, perhaps, they are planting the nobler oak-corn (acorn) in its place. In various parts of the country an army of grown-up boys assembles for a squirrel hunt. They choose sides, and the side that kills the greatest number of thousands enjoys a supper at the expense of the other side, and the whole neighborhood rejoices. Would it [not] be far more civilized and humane, not to say godlike, to recognize once in the year by some significant symbolical ceremony the part which the squirrel plays, the great service it performs, in the economy of the universe?

OCTOBER 22, 1860

In my boating of late I have several times scared up a couple of summer ducks of this year, bred in our meadows. They allowed me to come quite near, and helped to people the river. I have not seen them for some days. Would you know the end of our intercourse? Goodwin shot them, and Mrs. ———, who never sailed on the river, ate them. Of course, she knows not what she did. What if I should eat her canary? Thus we share each other's sins as well as burdens. The lady who watches admiringly the matador shares his deed. They belonged to me, as much as to any one, when they were alive, but it was considered of more importance that Mrs. ——— should taste the flavor of them dead than that I should enjoy the beauty of them alive.

AUGUST 16, 1858

In Brooks's barn I saw twenty-two gray squirrel skins freshly tacked up. He said that as many as one hundred and fifty had been killed this fall within a mile of his barn. They had been very numerous. His brother killed sixteen in one day a month ago.

DECEMBER 15, 1853

I see a brute with a gun in his hand, standing motionless over a musquash-house which he has destroyed. I find that he has visited every one in the neighborhood of Fair Haven Pond, above and below, and broken them all down, laying open the interior to the water, and then stood watchful, close by, for the poor creature to show its head there for a breath of air. There lies the red carcass of one whose pelt he has taken on the spot, flat on the bloody ice. And for his afternoon's cruelty that fellow will be rewarded with a ninepence, perchance. When I consider what are the opportunities of the civilized man for getting ninepences and getting light, this seems to me more savage than savages are. Depend on it that whoever thus treats the musquash's house, his refuge when the water if frozen thick, he and his family will not come to a good end. So many of these houses being broken open,—twenty or thirty I see . . .

DECEMBER 26, 1859

What a pitiful business is the fur trade, which has been pursued now for so many ages, for so many years by famous companies which enjoy a profitable monopoly and control a large portion of the earth's surface, unweariedly pursuing and ferreting out small animals by the aid of all the loafing class tempted by rum and money, that you may rob some little fellow-creature of its coat to adorn or thicken your own, that you may get a fashionable covering in which to hide your head, or a suitable robe in which to dispense justice to your fellowmen! Regarded from the philosopher's point of view, it is precisely on a level with rag and bone picking in the streets of the cities. The Indian led a more respectable life before he was tempted to debase himself so much by the white man. Think how many musquash and weasel skins the Hudson's Bay Company pile up annually in their warehouses, leaving the bare red carcasses on the banks of the streams throughout all British America,—and this it is, chiefly, which makes it *British* America. It is the place where Great Britain goes a-mousing. We have heard much of the wonderful intelligence of the beaver, but that regard for the beaver is all a pretense, and we would give more for a beaver hat than to preserve the intelligence of the whole race of beavers.

When we see men and boys spend their time shooting and trapping musquash and mink, we cannot but have a poorer opinion of them, unless we thought meanly of them before. Yet the world is imposed on by the fame of the Hudson's Bay and Northwest Fur Companies, who are only so many partners more or less in the same sort of business, with thousands of just such loafing men and boys in their service to abet them. On the one side is the Hudson's Bay Company, on the other the company of scavengers who clear the sewers of Paris of their vermin. There is a good excuse for smoking out or poisoning rats which infest the house, but when they are as far off as Hudson's Bay, I think that we had better let them alone. To such an extent do time and distance, and our imaginations, consecrate at last not only the most ordinary, but even vilest pursuits. The efforts of legislation from time to time to stem the torrent are significant as showing that there is some sense and conscience left, but they are insignificant in their effects. We will fine Abner if he shoots a singing bird, but encourage the army of Abners that compose the Hudson's Bay Company.

One of the most remarkable sources of profit opened to the Yankee within a year is the traffic in skunk-skins. I learn from the newspapers—as from other sources (*vide* Journal of Commerce in *Tribune* for April 5, 1859)—that "the traffic in skunk-skins has suddenly become a most important branch of the fur trade, and the skins of an animal which three years ago were deemed of no value whatever, are now in the greatest demand." "The principal markets are Russia and Turkey, though some are sent to Germany, where they are sold at a large profit." Furs to Russia! "The black skins are valued the most, and during the past winter the market price has been as high as one dollar per skin, while mottled skins brought only seventy cents." "Upward of 50,000 of these skins have been shipped from this city [New York] alone within the past two months." Many of them "are designed for the Leipsic sales, Leipsic being next to Novgorod, in Russia, the most important fur *entrepôt* in Europe. The first intimation received in this market of the value of this new description of fur came from the Hudson's Bay Company, which, having shipped a few to London at a venture, found the returns so profitable that they immediately prosecuted the business on an extensive scale." "The heaviest collections are made in the Middle and Eastern States, in some parts of which the mania for capturing these animals seems to have equalled the Western Pike's Peak gold excitement, men, women, and children turning out *en masse* for that purpose." And beside, "our fur dealers also receive a considerable sum for the *fat* of these animals!!"

Almost all smile, or otherwise express their contempt, when they hear of this or the rat-catching of Paris, but what is the difference between catching and skinning the skunk and the mink? It is only in the name. When you pass the palace of one of the managers of the Hudson's Bay Company, you are reminded that so much he got for his rat-skins. In such a snarl and contamination do we live that it is almost impossible to keep one's skirts clean. Our sugar and cotton are stolen from the slave, and if we jump out of the fire, it is wont to be into the frying-pan at least. It will not do to be thoughtless with regard to any of our valuables or property. When you get to Europe you will meet the most tender-hearted and delicately bred lady, perhaps the President of the Antislavery Society, or of that for the encouragement of humanity to animals, marching or presiding with the scales from a tortoise's back—obtained by laying live coals on it to make them curl up—stuck in her hair, rat-skin fitting as close to her fingers as erst to the rat, and, for her cloak, trimmings perchance adorned with the spoils of a hundred skunks,—rendered inodorous, we trust. Poor misguided woman! Could she not wear other armor in the war of humanity?

When a new country like North America is discovered, a few feeble efforts are made to Christianize the natives before they are all exterminated, but they are not found to pay, in any sense. But the energetic traders of the discovering country organize themselves, or rather inevitably crystallize, into a vast rat-catching society, tempt the natives to become mere vermin-hunters and rum-drinkers, reserving half a continent for the field of their labors. Savage meets savage, and the white man's only distinction is that he is the chief.

She says to the turtle basking on the shore of a distant isle, "I want your scales to adorn my head" (though fire be used to raise them); she whispers to the rats in the wall, "I want your skins to cover my delicate fingers;" and, meeting an army of a hundred skunks in her morning walk, she says, "worthless vermin, strip off your cloaks this instant, and let me have them to adorn my robe with;" and she comes home with her hands muffled in the pelt of a gray wolf that ventured abroad to find food for its young that day.

When the question of the protection of birds comes up, the legislatures regard only a low use and never a high use; the best-disposed legislators employ one [person], perchance, only to examine their crops [stomachs] and see how many grubs or cherries they contain, and never to study their dispositions, or the beauty of their plumage, or listen and report on the sweetness of their song. The legislature will preserve a bird professedly not because it is a beautiful creature, but because it is a good scavenger or the like. This, at least, is the defense set up. It is as if the question were whether some celebrated singer of the human race—some Jenny Lind or another—did more harm or good, should be destroyed, or not, and therefore a committee should be appointed, not to listen to her singing at all, but to examine the contents of her stomach and see if she devoured anything which was injurious to the farmers and gardeners, or which they cannot spare.

APRIL 8, 1859

The loon comes in the fall to sail and bathe in the pond, making the woods ring with its wild laughter in the early morning, at rumor of whose arrival all Concord sportsmen are on the alert, in gigs, on foot, two by two, three [by three], with patent rifles, patches, conical balls, spy-glass or open hole over the barrel. They seem already to hear the loon laugh; come rustling through the woods like October leaves, these on this side, those on that, for the poor loon cannot be omnipresent; if he dive here, must come up somewhere. The October wind rises, rustling the leaves, ruffling the pond water, so that no loon can be seen rippling the surface. Our sportsmen scour, sweep the pond with spy-glass in vain, making the woods ring with rude [?] charges of powder, for the loon went off in that morning rain with one loud, long, hearty laugh, and our sportsmen must beat a retreat to town and stable and daily routine, shop work, unfinished jobs again.

Or in the gray dawn the sleeper hears the long ducking gun explode over toward Goose Pond, and, hastening to the door, sees the remnant of a flock, black duck or teal, go whistling by with outstretched neck, with broken ranks, but in ranger order. And the silent hunter emerges into the carriage road with ruffled feathers at his belt, from the dark pond-side where he has lain in his bower since the stars went out.

C. 1845–1847

George Melvin, our Concord trapper, told me that in going to the spring near his house, where he kept his minnows for bait, he found that they were all gone, and immediately suspected that a mink had got them; so he removed the snow all around and laid open the trail of a mink underneath, which he traced to his hole, where were the fragments of his booty. There he set his trap, and baited it with fresh minnows. Going again soon to the spot, he found one of the mink's fore legs in the trap gnawed off near the body, and, having set it again, he caught the mink with his three legs, the fourth having only a short bare bone sticking out.

When I expressed some surprise at this, and said that I heard of such things but did not know whether to believe them, and was now glad to have the story confirmed, said he: "Oh, the muskrats are the greatest fellows to gnaw their legs off. Why, I caught one once that had just gnawed his third leg off, this being the third time he had been trapped; and he lay dead by the trap, for he couldn't run on one leg."

C. 1837–1847

When we started, saw some fishermen kindling their fire for spearing by the riverside. It was a lurid, reddish blaze, contrasting with the white light of the moon, with dense volumes of black smoke from the burning pitch pine roots rolling upward in the form of an inverted pyramid. The blaze reflected in the water, almost as distinct as the substance. It looked like tarring a ship on the shore of the Styx or Cocytus. For it is still and dark, notwithstanding the moon, and no sound but the crackling of the fire. The fishermen can be seen only near at hand, though their fire is visible far away; and then they appear as dusky, fuliginous figures, half enveloped in smoke, seen only by their enlightened sides. Like devils they look, clad in old coats to defend themselves from the fogs, one standing up forward holding the spear ready to dart, while the smoke and flames are blown in his face, the other paddling the boat slowly and silently along close to the shore with almost imperceptible motion.

OCTOBER 6, 1851

The spearer's light last night shone into my chamber on the wall and awakened me.

APRIL 8, 1853

Yesterday the hounds were heard. It was a hunter's day. All tracks were fresh, the snow deep and light. I met Melvin with his [game] bag full.

JANUARY 16, 1853

I see, in J. P. Brown's field, by Nut Meadow Brook, where a hen has been devoured by a hawk probably. The feathers whiten the ground. They cannot carry a large fowl very far from the farmyard, and when driven off are frequently baited and caught in a trap by the remainder of their quarry.

APRIL 6, 1853

The wood thrush Mr. Barnum never hired nor can, though he could bribe Jenny Lind and put her into his cage.

MAY 11, 1853

The inhumanity of science concerns me, as when I am tempted to kill a rare snake that I may ascertain its species. I feel that this is not the means of acquiring true knowledge.

MAY 28, 1854

Garfield says he found a hen-hawk's nest near Holden's Swamp (the old ones had got his chickens), sixty feet up a white pine. He climbed up and set a trap in it baited with a fish, with a string ten feet long attached. The young, but just hatched, faced him, and he caught the old one by the legs thus.

SEPTEMBER 8, 1854

Rice told his turtle story the other night: "One day I was going through Boston market and I saw a huddle of men around something or other. I edged my way between them and saw that they had got a great mud turtle on a plank, and a butcher stood over him with a cleaver in his hand. 'Eh,' said I, 'what are you trying to do?' 'We are waiting for him to put out his head so that we may cut it off. Look out,' they said; 'don't come so near, or he'll bite you.' 'Look here,' said I, 'let me try. I guess I can make him put his head out.' 'Let him try. Let him try,' they said, with a laugh. So I stepped into the ring and stood astride of the turtle, while they looked on to see the sport. After looking at him a moment, I put down my hands and turned him over on to his back, where-upon he immediately ran out his head and pushed against the plank to turn himself back, but, as they were not ready to cut at once, or his neck was not in a good position, I seized his head in both hands and, putting my feet against his breast-bone, drew his head out the full length of his neck and said, 'Now cut away. Only take care you don't cut my fingers.' They cut, and I threw the head down on the floor. As I walked away, some one said, 'I guess that fellow has seen mud turtles before to-day.'"

NOVEMBER 19, 1855

The Passenger Pigeon

> Saw pigeons in the woods, with their inquisitive necks and long tails, but few representatives of the great flocks that once broke down our forests.
>
> MAY 9, 1852

Thoreau lived at an unfortunately opportune time to document the historical decline and eventual extinction of the passenger pigeon. On the basis of the writings of earlier naturalists, discussions with his scientific acquaintances, and recollections of his older neighbors, Thoreau recognized that a massive decline had already occurred in the abundance of the pigeon and that what he saw in New England was only a vestige of the activity and impact that this species once had in the forests of eastern and central North America. He was familiar with early reports such as this one by Richard Hazen as he surveyed the northern boundary of Massachusetts between the Connecticut River and the Deerfield River in 1741: "For three miles together the [Passenger] Pigeons' nests were so thick that five hundred might have been told on the beech trees at one time; and could they have been counted on the hemlocks as well, I doubt not but five thousand, at one turn round." And Thoreau knew that such accounts were not isolated; he had read extensively the writings of naturalists such as Alexander Wilson and his descriptions of the "rivers" of pigeons in the upper Midwest: "A column, eight or ten miles in length, would appear . . . The leaders of this great body would sometimes gradually vary their course, until it formed a large bend of more than a mile in diameter, those behind tracing the exact route of their predecessors. This would continue sometimes long after both the extremities were beyond the reach of sight, so that the whole, with its glittering undulations, marked by a space on the face of the heavens resembling the winding of a vast and majestic river . . . Sometimes a hawk would make a swoop on a particular part of the column, from great height, when almost as quick as lightning, that part shot downwards out of the common track, but soon rising again, continued advancing at the same height as before; this inflection was continued by those behind, who on arriving at this point, dived down almost perpendicularly, to a great depth, and rising followed the exact path of those that went before" (*American Ornithology*, p. 232).

By Thoreau's time, the huge flocks that the colonists and early naturalists had witnessed were gone—the flocks that "broke down forests," dispersed acorns and chestnuts by the bushelful, filled the stews and pies of many households, and provided food for countless hungry hogs. But even in this period of declining flocks of birds Thoreau was able to document large numbers of passenger pigeons darting through the forest canopy and perching in the treetops, with the sunlight reflecting off their iridescent feathers. His journals also convey a sense of the unusual role of this now-extinct species in his cultural landscape.

The human activity generated by the arrival and passage of the pigeons on their annual migrations across New England was substantial. Each spring rural communities across the northeastern United States prepared for the birds' arrival, and the sight of the first advance birds was always noteworthy and much anticipated. The squab (an unfledged pigeon) and older birds were consumed in every rural household and were also packed and shipped by the barrelful to the cities and out of most eastern ports. Other parts of the birds were put to a variety of uses—feathers for stuffing and insulation and bones by the millions for fertilizer.

The huge numbers of birds and their habit of roosting by the thousands in treetops where they built crude nests of stout twigs resulted in extensive mortality of juveniles and adults alike as the sheer weight of birds broke branches and caused trees to fall. This natural mortality, coupled with killing by farmers and professional pigeon hunters, left carcasses strewn through the forests and led to the common practice of releasing the hogs into the woods to feed on dead birds as the live multitudes frolicked above. Many farmers kept a few pigeons in cages to bring out during migration when they would be strung up on poles or pinned flapping on the ground, serving as live decoys to attract their wild relatives. The migrating flocks were also enticed with corn, other grains, or ground acorns to lure them groundward as they passed overhead. Large treetop platforms and hillside stands were erected from which the birds could be netted or shot, and artificial perches were established in roosting areas where the pigeons concentrated and could be easily beaten down by the dozen with poles. In all of Thoreau's descriptions of this activity, the declining status of the passenger pigeon is apparent.

Although Thoreau does not dwell on the passenger pigeon in his writings, his journal entries lead us to consider what we have lost through the disappearance of this species and what that loss means to the environment of eastern forests. Thoreau mentions two important impacts of passenger pigeons: first, they helped to keep the land and forests in a state of continual flux because their tendency to roost in flocks of thousands resulted in the shredding of branches, the breakage of trees, and occasional damage to entire forest stands. Second, passenger pigeons played a critical role in the ongoing process of spreading plant populations across the landscape. These birds had a superb capacity to disperse seeds by virtue of their phenomenal numbers and size, prodigious appetites, and rapid and lengthy flights. The sight, sound, and ecological impact of endless streams of pigeons are difficult to comprehend today, but typical North American accounts from the seventeenth and eighteenth centuries describe flocks consisting of hundreds of thousands to many millions of birds that required hours to stream by overhead. This habit of flocking together and concentrating in dense roosts contributed greatly to the decline of the species because it made them easy to hunt and trap. It also made them a powerful ecological force as a result of breakage of trees and damage to forests. Thoreau refers frequently to this breakage of forests, which made a lasting impression on his contemporary John James Audubon (1840): "The pigeons arriving by thousands, alighted everywhere, one above another, until solid masses were formed on the branches all round. Here and there the perches gave way under the weight with a crash, and, falling to the ground, destroyed hundreds of birds beneath, forcing down the dense group with which every stick was loaded."

As a boy in Wisconsin, John Muir actively explored the consequences of this environmental impact wrought by the passenger pigeons. Woken in the night by the overpowering sound of branch and tree breakage, as though by a major windstorm, Muir surveyed the scene the next day to discover a severely damaged forest and dozens of dead pigeons, killed by falling branches and strewn across the ground. In his journals Muir remarked that tree species must vary considerably in their ability to withstand such pigeon roosting and damage, and he went on to speculate that in the areas where pigeons congregate this differential impact might encourage the stout, resistant trees such as oaks to predominate over species with weaker branches, such as red maple. Similarly, eighteenth-century descriptions of pigeon dung covering a hundred acres of forest to a depth of inches in nesting areas suggest that the birds served a little appreciated role in maintaining forest nutrients. Consequently, the pigeon was an ecological force to be reckoned with—similar to wind, fire, or logging in its ability to affect the structure and composition of the forest.

Thoreau observed the feeding behavior of pigeons and reported that analyses of the stomach contents of dead pigeons by his acquaintances confirmed that they consumed acorns, chestnuts, beechnuts, many types of berries, and white pine seed, along with a diverse assortment of cultivated and natural grains. Thoreau was extremely impressed by the ease

with which pigeons could ingest, carry, and disperse seeds over tens to hundreds of miles, and he used this image to trivialize human tendencies to establish (and protect) property lines and political boundaries. As highlighted in his writings on succession, he also clearly recognized that seed dispersal by animals such as pigeons plays a critical element in the dynamics of natural landscapes. Season after season millions of birds would undertake a grand natural experiment in which seeds were carried into new and distant locations, where they had the chance of germinating and growing if the conditions proved right. Through this process trees were replaced as they died, forests could regenerate after natural or human disturbance, and a species could extend its existence if the broader environment should change. Interestingly, Thoreau provides little evidence that the pigeons had an impact on crops, which gives support to the view that the birds' effect on farms or other human enterprises had little to do with the slaughter that was inflicted on the species.

Human greed was not the only factor that led to the decline of the species. The passenger pigeon's habit of producing just a single egg in a flimsy nest of twigs greatly limited its potential rate of reproduction. The birds were further threatened by the ongoing reduction in forest land, which fragmented the natural habitat of food, nest sites, and roosts.

What has been the consequence of eliminating these enormous flocks of birds from the landscape? Our ecosystems are remarkably resilient to such losses, in the sense that forests persist and still contain most of the plants and animals that must have coexisted with the pigeon flocks. But the disappearance of such a major force in nature cannot come without extensive change and readjustment. Perhaps, like the infrequent hurricane, heavy ice storm, or fire, the physical disruption caused by these birds was important in generating a continual mosaic of change and thereby diversified the structure and composition of forests. It is likely that the enigmatically rapid rate of migration of many tree species into the northeastern United States from southern unglaciated areas after the last Ice Age is partly due to the effectiveness of the pigeons and other animal species in transporting seeds; if so, the pigeons may play a critical role in nature's adjustment to rapid environmental change. Undoubtedly the passenger pigeon was always important in the ongoing movement of a variety of plants across the landscape, and in such numbers it must have been a major food source for many larger birds and mammals. Speculating on the changes resulting from the loss of this species can make us aware of the extent of the web of effects that humans exert on the landscape. Although we tend to focus our attention on the major, direct impacts of human land use and environmental change, many indirect consequences of our activities are equally important, though subtle and unseen.

There are, of course, other roles played by plants and animals in our environment that influence the way it looks and sounds, roles that shape its seasonal quality and change. Our sense of the environment and its seasonal changes may be very different today in the absence of a bird that once constituted up to 10 percent of the total number of birds in existence on earth. As Thoreau did, we may read descriptions such as those by Alexander Wilson and realize that the pigeon was not just an ecological force but was very much a part of the environment.

Thoreau lived in the period of the most rapid decline of the species, as the new systems of telegraphs and railroads allowed hunters to keep up with the flocks and to supply distant markets with their product. As a consequence, the passenger pigeon was eliminated from the Massachusetts landscape by 1879, less than two decades after Thoreau wrote his last notes recording its activity. The last member of the species in geological history, a female pigeon by the name of Martha, died in captivity at the Cincinnati Zoo on September 1, 1914.

The pigeon carries an acorn in his crop from the King of Holland's to Mason and Dixon's line. Yet we think if rail fences are pulled down and stone walls set up on our farms, bounds are henceforth set to our lives and our fates decided.

MARCH 21, 1840

Saw a pigeon-place on George Heywood's cleared lot,—the six dead trees set up for the pigeons to alight on, and the brush house close by to conceal the man. I was rather startled to find such a thing going now in Concord. The pigeons on the trees looked like fabulous birds with their long tails and their pointed breasts. I could hardly believe they were alive and not some wooden birds used for decoys, they sat so still; and, even when they moved their necks, I thought it was the effect of art. As they were not catching then, I approached and scared away a dozen birds who were perched on the trees, and found that they were freshly baited there, though the net was carried away, perchance to some other bed. The smooth sandy bed was covered with buckwheat, wheat or rye, and acorns. Sometimes they use corn, shaved off the ear in its present state with a knife. There were left the sticks with which they fastened the nets. As I stood there, I heard a rushing sound, and looking up, saw a flock of thirty or forty pigeons dashing toward the *trees,* who suddenly whirled on seeing me and circled round and made a new dash toward the bed, as if they would fain alight if I had not been there, then steered off. I crawled into the bough house and lay awhile looking through the leaves, hoping to see them come again and feed, but they did not while I stayed. This net and bed belong to one Harrington of Weston, as I hear. Several men still take pigeons in Concord every year; by a method, methinks, extremely old and which I seem to have seen pictured in some old book of fables or symbols, and yet few in Concord know exactly how it is done. And yet it is all done for money and because the birds fetch a good price, just as the farmers raise corn and potatoes. I am always expecting that those engaged in such a pursuit will be somewhat less grovelling and mercenary than the regular trader or farmer, but I fear that it is not so.

SEPTEMBER 12, 1851

He had ten live pigeons in a cage under his barn. He used them to attract others in the spring. The reflections from their necks were very beautiful. They made me think of shells cast up on a beach. He placed them in a cage on the bed and could hear them prate at the house.

The turtle doves plagued him, for they were restless and frightened the pigeons.

DECEMBER 15, 1853

Brooks has let out some of his pigeons, which stay about the stands or perches to bait others. Wild ones nest in his woods quite often. He begins to catch them the middle of August.

JULY 18, 1854

I scare pigeons from Hubbard's oaks beyond. How like the creaking of trees the slight sounds they make! Thus, they are concealed. Not only their *prating* or *quivet* is like a sharp creak, but I heard a sound from them like a dull grating or cracking of bough on bough.

SEPTEMBER 12, 1854

On a white oak beyond Everett's orchard by the road, I see quite a flock of pigeons; their blue-black droppings and their feathers spot the road. The bare limbs of the oak apparently attracted them, though its acorns are thick on the ground. These are found whole in their crops. They swallow them whole. I should think from the droppings that they had been eating berries. I hear that Wetherbee caught ninety-two dozen last week.

SEPTEMBER 12, 1854

Dense flocks of pigeons hurry-skurry over the hill. Pass near Brooks's pigeon-stands. There was a flock perched on his poles, and they sat so still and in such regular order there, being also the color of the wood, that I thought they were wooden figures at first.

SEPTEMBER 15, 1859

I sat near Coombs's pigeon-place by White Pond. The pigeons sat motionless on his bare perches, from time to time dropping down into the bed and uttering a *quivet* or two. Some stood on the perch; others squatted flat. I could see their dove-colored breasts. Then all at once, being alarmed, would take flight, but ere long return in straggling parties. He tells me that he has fifteen dozen baited, but does not intend to catch any more at present, or for two or three weeks, hoping to attract others. Rice says that white oak acorns pounded up, shells and all, make the best bait for them.

SEPTEMBER 21, 1859

The white pine seed is very abundant this year, and this must attract more pigeons. Coombs tells me that he finds the seed in their crops. Also that he found within a day or two a full-formed egg with shell in one.

SEPTEMBER 28, 1859

In the meanwhile we came upon another pigeon-bed, where the pigeons were being baited, a little corn, etc., being spread on the ground, and, [as?] at the first, the bower was already erected.

AUGUST 15, 1854

Standing by a pigeon-place on the north edge [of] Damon's lot, I saw on the dead top of a white pine four or five rods off—which had been stripped for fifteen feet downward that it might die and afford with its branches a perch for the pigeons about the place, like the more artificial ones that were set up—two woodpeckers that were new to me.

OCTOBER 8, 1860

The woods on the neighboring shore were alive with pigeons, which were moving south, looking for mast, but now, like ourselves, spending their noon in the shade. We could hear the slight, wiry, winnowing sound of their wings as they changed their roosts from time to time, and their gentle and tremulous cooing. They sojourned with us during the noontide, greater travellers far than we. You may frequently discover a single pair sitting upon the lower branches of the whitepine in the depths of the wood, at this hour of the day, so silent and solitary, and with such a hermit-like appearance, as if they had never strayed beyond its skirts, while the acorn which was gathered in the forests of Maine is still undigested in their crops. We obtained one of these handsome birds, which lingered too long upon its perch, and plucked and broiled it here . . . for our supper.

A WEEK ON THE CONCORD AND MERRIMACK RIVERS (1839)

The American Chestnut

A squirrel goes a-chest-nutting perhaps as far as the boys do, and when he gets there he does not have to shake or club the tree or wait for frost to open the burs; he walks[?] up to the bur and cuts it off, and strews the ground with them before they have opened. And the fewer they are in the wood the more certain it is that he will appropriate every one, for it is no transient afternoon's picnic with him, but the pursuit of his life, a harvest that he gets as surely as the farmer his corn.

Now it is important that the owners of these wood-lots should know what is going on here and treat them and the squirrels accordingly.

OCTOBER 17, 1860

Within fifty years of Thoreau's death, the chestnut blight had been introduced into the Brooklyn Botanical Garden via Asiatic chestnut trees and had spread to Concord, Massachusetts. There, and rapidly across the natural range of the species, the blight killed all mature trees, leaving a legacy of slowly decomposing dead snags and stumps scattered throughout our forests and countless numbers of thriving sprouts that grow rapidly to a size of 20 to 30 feet in height and 2 to 6 inches in diameter before they are reinfected by the fungal disease and killed back to the roots. Thus the story of the chestnut is a classic tale of humans interacting with nature to shape and move it in novel directions. This tree, once a very abundant species that had played a uniquely important role in the ecology of eastern forests for millennia and provided a phenomenal array of products for an expanding human population, has been eliminated (as a full-grown tree) from the land. Tracing its history reminds us again that the cultural landscape we occupy embraces not only the landscape elements that we have created, but also those natural features that we have removed.

What do we actually know of the chestnut's former place in our forests and culture? We do have many historical photographs of grand chestnut trees and the forests in which they grew. Because the chestnut was so important in construction, we also can find massive beams and countless boards cut from chestnut logs in bridges, barns, and houses. Graying stumps and fallen trunks of the trees lie scattered in forests across the eastern United States more than eighty years after their death, attesting to the wood's remarkable resistance to decay and hinting at the chestnut's former abundance. And we can still see the widespread and numerous sprouts, which retain their long yellow leaves into the fall to lend intrigue to the question of chestnut's original role in the woods.

In Thoreau's journal entries we receive an evenhanded and far-reaching account of the place that the chestnut tree occupied in the forests and the daily life of nineteenth-century New England. Unlike any interpretation we could develop today, Thoreau's description is a record unbiased by knowledge of the blight to come. Rather, it is a day-to-day account of a species so common that it would have been taken completely for granted had it not been so important to human and forest life in a variety of ways. Because chestnut was a food, a building material, a fuel, and a remarkable example of natural history, Thoreau devoted more space in his journal to this one tree species than to almost any other subject.

It is the small details that Thoreau provides about the chestnut that first capture one's attention. The abundance of chestnut leaves on the ground created a constant rustle in the autumn woods and generated a loud sound as one walked through them. Thoreau likened this distinctive noise to the tumult of waves dashing against each other as he took an autumn sail down a forest path, brushing the leaves aside in his wake. The leaves concealed the fallen nuts, and on this subject of chest*nuts* we receive a great amount of detail from the journals. The chestnuts ripened in October, were often brought down by severe frost, and then were covered by the falling leaves, which were loosened by wind and rain. The profusion of nuts was remarkable; they rained down from the trees to cover the ground and fill the wagon ruts on the road.

The autumn ripening of this harvest was accompanied by great activity from both wildlife and man. The jays screamed, squirrels scolded, and mice scurried through the leaves to recover the valuable food. The human sounds that filled the woods were even more remarkable: there were resounding echoes as children and adults beat the nut-laden trees with clubs and stones in order to dislodge the chestnuts, followed by shouts of glee as the booty fell into the collecting blankets. Today it is difficult to imagine this primeval sound of tree-clubbing in our woods. The annual bruising resulting from the nut collection was severe enough that the trees bore scars up and down their trunks and branches. When not consumed, the nuts were hoarded by the birds and mammals; once dispersed, they produced seedlings that Thoreau located up to half a mile away from the nearest tree.

Chestnut trees were cut for diverse products: lumber, rails, poles, posts, railroad sleepers, and coffins, to name a few. The wood's value was so great and the harvesting was so rampant that Thoreau feared the species was declining from overuse. Consequently, he argued that an understanding of factors such as the chestnut's seed dispersal, seedling growth, and sprouting rate could be used to manage it wisely and thereby reverse this trend. Chestnut's modern reputation of having an unsurpassed ability to sprout and grow rapidly is affirmed in these discussions. Thoreau notes that in one year chestnut sprouts typically reached six feet in height and exceeded an inch in diameter. He also regularly recorded sustained growth of nearly one-half inch in diameter per year in trees ranging from four inches to four feet across, truly a remarkable rate of growth. When he counted the growth rings on stumps up to four and a half feet in diameter, he learned that it took only 75 to 80 years to reach this massive size. Forest-grown trees were tall and narrow with few lower branches. But in open land at the edge of a field or out in a pasture, chestnuts would get immense, reaching upward and outward with a huge broad crown and widely spreading low branches. The largest chestnut tree Thoreau encountered was a full seven feet in diameter, evidence of the size and age (older than 150 years) that this species could reach.

The chestnut was a remarkable and distinctive plant. It played a crucial

HOMER

role in the economy and activity of southern New England farms, households, and towns, and it provided a regular and abundant source of food for a wide range of animals. As we walk through our own forests today, we may assume that they are natural and intact. However, having read through these journals, one may rightly question what impact the loss of a single species may have on a forest ecosystem and cultural landscape. We can see that the oaks, maples, and birches have benefited from the chestnut's new status as a forest shrub. The forests may appear intact, but a host of other species and natural processes have changed as this once great tree has diminished in stature.

The chestnut leaves already rustle with a great noise as you walk through the woods, as they lie light, firm, and crisp. Now the chestnuts are rattling out. The burs are gaping and showing the plump nuts. They fill the ruts in the road, and are abundant amid the fallen leaves in the midst of the wood. The jays scream, and the red squirrels scold, while you are clubbing and shaking the trees. Now it is true autumn; all things are crisp and ripe.

OCTOBER 11, 1852

We make a great noise going through the fallen leaves in the woods and wood-paths now, so that we cannot hear other sounds, as of birds or other people. It reminds me of the tumult of the waves dashing against each other or your boat. This is the dash we hear as we sail the woods.

OCTOBER 28, 1860

The rain of the night and morning, together with the wind, had strewn the ground with chestnuts. The burs, generally empty, come down with a loud sound, while I am picking the nuts in the woods. I have come out before the rain is fairly over, before there are any fresh tracks on the Lincoln road by Britton's shanty, and I find the nuts abundant in the road itself. It is a pleasure to detect them in the woods amid the firm, crispy, crackling chestnut leaves. There is somewhat singularly refreshing in the color of this nut, the chestnut color. No wonder it gives a name to a color. One man tells me he has bought a wood-lot in Hollis to cut, and has let out the picking of the chestnuts to women at the halves. As the trees will probably be cut for them, they will make rapid work of it.

OCTOBER 15, 1852

I get a couple of quarts of chestnuts by patiently brushing the thick beds of leaves aside with my hand in successive concentric circles till I reach the trunk; more than half under one tree. I believe I get more by resolving, where they are reasonably thick, to pick all under one tree first. Begin at the tree and brush the leaves with your right hand in toward the stump, while your left holds the basket, and so go round and round it in concentric circles, each time laying bare about two feet in width, till you get as far as the boughs extend. You may presume that you have got about all then. It is best to reduce it to a system. Of course you will shake the tree first, if there are any on it. The nuts lie commonly two or three together, as they fell.

OCTOBER 24, 1857

I was this afternoon gathering chestnuts at Saw Mill Brook. I have within a few weeks spent some hours thus, scraping away the leaves with my hands and feet over some square rods, and have at least learned how chestnuts are planted and new forests raised. First fall the chestnuts with the severe frosts, the greater part of them at least, and then, at length, the rains and winds bring down the leaves which cover them with a thick coat. I have wondered sometimes how the nuts got planted which merely fell on to the surface of the earth, but already I find the nuts of the present year partially mixed with the mould, as it were, under the decaying and mouldy leaves, where is all the moisture and manure they want. A large proportion of this year's nuts are now covered loosely an inch deep under mouldy leaves, though they are themselves sound, and are moreover concealed from squirrels thus.

DECEMBER 31, 1852

Now is the time for chestnuts. A stone cast against the trees shakes them down in showers upon one's head and shoulders. But I cannot excuse myself for using the stone. It is not innocent, it is not just, so to maltreat the tree that feeds us. I am not disturbed by considering that if I thus shorten its life I shall not enjoy its fruit so long, but am prompted to a more innocent course by motives purely of humanity. I sympathize with the tree, yet I heaved a big stone against the trunks like a robber,—not too good to commit murder. I trust that I shall never do it again.

OCTOBER 23, 1855

I hear the dull thump of heavy stones against the trees from far through the rustling wood, where boys are ranging for nuts.

OCTOBER 24, 1857

I see where the chestnut trees have been sadly bruised by the large stones cast against them in previous years and which still lie around.

OCTOBER 18, 1856

Young Weston said that they found, in redeeming a meadow, heaps of chestnuts under the grass, fifteen rods from the trees, without marks of teeth. Probably it was the work of the meadow mice.

DECEMBER 18, 1853

Went a-chestnutting this afternoon to Smith's wood-lot near the Turnpike. Carried four ladies. I raked. We got six and a half quarts, the ground being bare and the leaves not frozen. The fourth remarkably mild day. I found thirty-five chestnuts in a little pile under the end of a stick under the leaves, near—within a foot of—what I should call a gallery of a meadow mouse.

JANUARY 10, 1853

Looking through this wood and seeking very carefully for oak seedlings and anything else of the kind, I am surprised to see where the wood was chiefly oak a cluster of little chestnuts six inches high and close together. Working my hand underneath, I easily lift them up with all their roots,—four little chestnuts two years old, which partially died down the first year,—and to my surprise I find still attached four great chestnuts from which they sprang and four acorns which have also sent up puny little trees beneath the chestnuts. These eight nuts all lay within a diameter of two inches about an inch and a half beneath the present leafy surface, in a very loose soil of but [?] half decayed leaves in the midst of this young oak wood. If I had not been looking for something of the kind, I should never have seen either the oaks or the chestnuts. Such is the difference between looking for a thing and waiting for it to attract your attention. In the last case you will probably never see it [at] all. They were evidently planted there two or three years ago by a squirrel or mouse. I was surprised at the sight of these chestnuts, for there are not *to my knowledge* any chestnut trees—none, at least, nearly large enough to bear nuts—within about half a mile of that spot, and I should about as soon have expected to find chestnuts in the artificial pine grove in my yard. The chestnut trees old enough to bear fruit are near the Lincoln line about half a mile east of this through the woods and over hill and dale. No one acquainted with these woods—not the proprietor—would have believed that a chestnut lay under the leaves in that wood or within a quarter of a mile of it, and yet from what I saw then and afterward I have no doubt that there were hundreds, which were placed there by quadrupeds and birds.

OCTOBER 17, 1860

It is well known that the chestnut timber of this vicinity has rapidly disappeared within fifteen years, having been used for railroad sleepers, for rails, and for planks, so that there is danger that this part of our forest will become extinct.

The last chestnut tracts of any size were on the side of Lincoln. As I advanced further through the woods toward Lincoln, I was surprised to see how many little chestnuts there were, mostly two or three years old and some even ten feet high, scattered through them and also under the dense pines, as oaks are. I should say there was one every half-dozen rods, made more distinct by their yellow leaves on the brown ground, which surprised me because I had not attended to the spread of the chestnut, and it is certain that every one of these came from a chestnut placed there by a quadruped or bird which had brought it from further east, where alone it grew.

OCTOBER 17, 1860

Where large chestnuts were sawed down last winter by Walden, sprouts have come up six feet high on every side of the stump, very thick, so as to form perfect bowers in which a man might be concealed.

OCTOBER 25, 1852

Measured a chestnut stump on Asa White's land, twenty-three and nine twelfths feet in circumference, eight and one half feet one way, seven feet the other, at one foot from ground.

JUNE 2, 1852

In Stow's wood at Saw Mill Brook an old chestnut stump. Two sprouts from this were cut three years ago and have forty-two rings. From the stumps of the sprouts, other sprouts three years old have grown. The old stump was cut there forty-five years ago. The centre of the stumps of each of these sprouts is hollow for one and a half inches in diameter. See a chestnut stump, a seedling sawed off, with seventy-five rings and no sprout from it. Commonly the sprouts stand in a circle around the stump,—often a dozen or more of them.

DECEMBER 2, 1860

I have seen more chestnuts in the streets of New York than anywhere else this year, large and plump ones, roasting in the street, roasting and popping on the steps of banks and exchanges. Was surprised to see that the citizens made as much of the nuts of the wild-wood as the squirrels. Not only the country boys, all New York goes a-nutting. Chestnuts for cabmen and newsboys, for not only are squirrels to be fed.

DECEMBER 1, 1856

The chestnut, with its tough shell, looks as if it were able to protect itself, but see how tenderly it has been reared in its cradle before its green and tender skin hardened into a shell. The October air comes in, as I have said, and the light too, and proceed to paint the nuts that clear, handsome reddish (?) brown which we call chestnut. Nowadays the brush that paints chestnuts is very active. It is entering into every open bur over the stretching forests' tops for hundreds of miles, without horse or ladder, and putting on rapid coats of this wholesome color. Otherwise the boys would not think they had got perfect nuts. And that this may be further protected, perchance, both within the bur and afterward, the nuts themselves are partly covered toward the top, where they are first exposed, with that same soft velvety down. And then Nature drops it on the rustling leaves, a *done* nut, prepared to begin a chestnut's course again. Within itself, again, each individual nut is lined with a reddish velvet, as if to preserve the seed from jar and injury in falling and, perchance, from sudden damp and cold, and, within that, a thin white skin enwraps the germ. Thus it is lining within lining and unwearied care,—not to count closely, six coverings at least before you reach the contents!

OCTOBER 22, 1857

Stepping Back and Looking Ahead

Reading Forest and Landscape History

> Our wood-lots, of course, have a history, and we may often recover it for a hundred years back, though we *do* not . . . Yet if we attended more to the history of our lots we should manage them more wisely.
>
> OCTOBER 16, 1860

In his quest to understand the landscape that surrounded him, Thoreau became a historian of nature, using a diverse range of clues in the natural world as well as human records to reconstruct landscape and forest history. He then explored the link between these earlier environments and his own world in order to develop a deeper appreciation of the forces that shaped nature and the sights that he encountered on his daily walks. From these historical pursuits by Thoreau we can, in turn, glean many interesting ecological insights and learn some new methods to employ in our own efforts to interpret forest and landscape change. A look backward at the landscape of the Indians and the early colonists provides a benchmark against which we can gauge the extent of change in nature that Thoreau subsequently described and that we now experience.

The motivations underlying historical reconstructions of nature are diverse. In the case of Thoreau, the enjoyment that comes from puzzling through and piecing together ecological and human history is evident and radiates from every volume of his writings. Daily, he would pose questions for himself concerning the origins of particular tree shapes, forest types, or landscape features that he encountered. Subsequently he would revisit these subjects over the following days or weeks, recording his emerging thoughts as he gathered additional evidence and ideas. The interpretation of changes in natural history and the reconstruction of the land-use activity of Indian cultures and early colonial settlers provided him with a simple, engaging, and endless pastime.

Scientific curiosity was another force driving Thoreau's historical studies. Thoreau was aware of the writings and emerging theories concerning evolution and glacial history coming from the work of Charles Darwin, the great British naturalist, and Louis Agassiz, the eminent Swiss geologist who presided over Harvard University's Museum of Comparative Zoology, and he was eager to integrate geological time scales and long-term processes into his understanding of the abundance and characteristics of a plant or animal or the appearance of a landscape. The explanation for present conditions often lies in a past that can be ferreted out from evidence on the site, in the plants and animals that inhabit an area, or in anecdotal descriptions contained in a deed or history book. Thus, by examining the history of a woodlot, puzzling through some dusty records in the town hall, or quizzing his acquaintances Therien or Minott on changes in land-use practices, Thoreau was often able to explain the current distribution of particular plants, the abundance and behavior of many animal species, or the ages and sequence of establishment of oaks and pines in a forest stand.

Nostalgia and a broadly romantic attitude toward nature were other factors contributing to Thoreau's interest in the past; he enjoyed searching for the remnants of primitive nature and the legacies of earlier landscapes and people that gave his own landscape a deeper and more intriguing history. In particular, he delighted in finding sights and forest stands that he sensed were closer in character to the "wildwood" or original forest conditions of New England. He searched for these areas locally, for example in the swamps of Concord, in the majestic old forest of huge oaks at Inches Wood in Boxboro, across north-central Massachusetts, and up into the Maine woods. He also sought out ancient forests and landscapes historically through the early writings of individuals such as William Wood, the British chronicler of the Massachusetts Bay Colony. Thoreau reveled in the wildness of these historical images and in the roughness of the few "natural" landscapes that he could locate. He then used these examples of primitive forests and raw nature to gauge the extent to which humans had tamed the New England landscape and as a source of information on how plants, animals, and the environment functioned naturally.

Thus, by studying areas in which human impact had been slight, Thoreau felt that he could see and understand processes that were absent or modified in the heavily cultivated and inhabited areas that composed most of the land.

The insights that Thoreau acquired through his studies of forest history and natural change have great relevance to modern-day natural history and conservation. By studying forests that had been changed very little by human activity, Thoreau was able to learn where particular tree species grew best, how long they might live, how tall they might grow, and what range of growth forms they might take. Through comparison of wildlife in old forests with that from the cut-over woodlots of nineteenth-century Massachusetts, he discovered which animal species seemed most sensitive to the cutting that was widespread in the land. This information is directly relevant to us today as we try to understand how our own landscape has developed and the ways in which our second-growth forests may change as they grow in size and age. By following the consequences of different types of human impact on nature, Thoreau also developed ideas for managing forests more effectively in order to yield more wood and conserve natural qualities. These practical observations by Thoreau are wide-ranging and suggest, for example, that the thinning of forests through selective logging will concentrate much greater growth in the remaining trees, that different species and sizes of trees vary significantly in their ability and rate of sprouting and regrowth after cutting or fire, and that the regeneration or development of a new forest differs tremendously on lands subjected to contrasting histories of human and natural disturbance.

From his reconstruction of forest change came one of Thoreau's most significant contributions to natural history: the concept of succession. Equally valuable observations and recommendations came in the areas of conservation and preservation. Thoreau proposed a long-term plan for the conservation of woodlots through ongoing, careful management that would ensure a supply of fuelwood into the future. Related observations led to his proposal that the few remaining wild and less developed areas be preserved in every township. In his native Concord, he specifically identified some diverse swamps and bogs, a handful of older woods, and the extensive Easterbrook area of northern Concord and the adjoining town of Carlisle. Interestingly, this last area (known locally today as the Estabrook woods) has been the focus of a far-reaching and very successful twentieth-century effort in land preservation that was based in large part on the natural, historical, and archaeological qualities identified by Thoreau.

The conclusions drawn from some of his historical reconstructions lent conviction to many of the internal debates that Thoreau held on such scientific issues as evolution, spontaneous generation, and plant migration. The observation of pine pollen along lakeshores, many hundreds of yards from the nearest pine tree, demonstrated to Thoreau the remarkable wind-dispersal capacity of some pollen. This, in turn, helped him to interpret the pollination and development of female cones on a small, solitary pine that grew in his front yard but was completely lacking in male cones. He argued that one did not need to invoke spontaneous generation, as many of his contemporaries did, in order to explain such cone development or to understand the arrival of small oaks in a pine forest, or pine and birch seedlings in an open field. Rather, he reasoned, much as pollen was lofted afar by breezes, squirrels carried the acorns into the pine wood and wind transported the pine and birch seed great distances into open fields. A careful examination of these locales generally revealed many such seeds and seedlings and contributed to Thoreau's understanding of landscape change. Eventually, the field would become pine and birch forest, and the pine woods, when cut, would develop into an oak forest that would retain mute evidence of its past in axe-scarred pine stumps, still visible even fifty years later.

Finally, in the physical records of forest history, Thoreau located evidence for some rare and infrequent natural events as well as for the patterns and consequences of prehistoric human activity. This information helped him understand the environmental forces that shaped his woods and offered a new perspective on the historical role of different plant and animal species. In subtle mounds of soil, often paired with an adjoining depression, he recognized the impact, and deduced the direction, of ancient windstorms that had uprooted an earlier generation of forest. Long after the uprooted trunk had decomposed or was carted off to fill some farmer's fireplace, the location and orientation of the upturned roots remained apparent in the microtopography of the forest floor. This evidence of ancient windthrow showed Thoreau that the particular site had been continuously forested for many decades or centuries, and was not a new forest developed recently and secondarily on an old pasture or plowed field. In similar fashion, subtle trails worn into the shores above Walden and the other Concord ponds and acres of relict cornhill mounds near Bateman's Pond were the legacies of the Indian civilization that had previously occupied Concord. For Thoreau, such clear evidence of Indian clearings for villages and corn fields, coupled with the wealth of arrowheads and artifacts that he collected and the charcoal that he noted in soils across the land, corroborated the early colonial descriptions of the openness of the original forests and the scattering of fields and meadow areas created through Indian burning and clearing. Regardless of the absolute numbers of Indians, if they were equipped with a knowledge of the use of fire they could manage and change the landscape quite effectively. Awareness of the range of Indian activity also suggested to Thoreau an earlier pattern for the wind-dispersed seeds of pine and birch. These species, he speculated, could have established in the villages, fields, and fire-created openings of early dwellers, much as they sprang up on neglected farmland and along fencerows in his day.

From the shape and structure of individual plants Thoreau gleaned considerable information on forest growth and history. Tall, straight trees supporting few low branches obviously grew up in a closed forest, where the individual trees shot upward to reach for the full sunlight above; Thoreau found these trees in old "primitive forests" (that is, forests that had never been cleared), where they suggested to him the characteristics of the wildwood that once occupied much of prehistoric Concord. Large, rounded, and spreading trees, with great branches extending horizontally near the ground, established and grew under open, sunlit conditions, perhaps as widely spaced shade trees in a pasture, along a stone wall, or as the first seedlings to establish in a recently abandoned field. A combination of these two growth forms indicated the presence of two generations of trees: an older, open-grown generation with broad branches surrounded by densely packed progeny.

The seedlings, saplings, and sprouts in the understory of a forest were often noted by Thoreau to be different species from the large, mature trees in the overstory. This generational shift in tree species was an indication to him of succession and the ongoing changes that can occur in the forest environment. The overstory canopy trees, perhaps pitch pine or white pine, had established in an abandoned field, where these species thrive in open sunlight and dry conditions and compete very effectively with dense grass. However, the pines, red cedar, and other "pioneers" are intolerant of shade, which prevents them from continuing to establish in the closed forest and leads to the establishment of a second generation of seedlings and saplings composed of more shade-tolerant species, perhaps oak, maple, beech, or hemlock. The herbs and smaller plants yield somewhat different clues. Old forests have particular plants and animals, perhaps because of their specific environment or as a consequence of the long continuity of forested conditions. In contrast, abundant grass, *Cladonia* lichens ("reindeer moss" and "British soldiers"), or an errant weed species revealed to Thoreau the secondary origins of a forest that had established following the abandonment of a pasture or old field.

Studies of the age and growth rates of individual plants enabled Thoreau to examine forest history quantitatively and quite precisely. On the stumps resulting from the fuelwood cutting that occurred regularly throughout Concord, the growth rings laid down annually by each tree provided him with a means to determine its age and year of establishment. From changes in the width of the individual growth rings through time, it was possible to interpret shifts in tree vigor that resulted from changes in the broader environment, including climate; from human impacts such as logging or fire; or from the gradual aging of the trees. Interestingly, Thoreau recognized a basic fact that is now well incorporated into all modern ecological studies of tree growth: namely, that as trees enlarge in diameter, a constant rate of wood production yields a decreasing rate of diameter increase through time. Thus, in order to correct for the inherent tendency of trees growing in a constant environment to enlarge less each year, Thoreau noted the measurements of tree growth in terms of both the diameter increase and the annual change in total wood production when he was studying the history of a forest.

Even smaller plants reveal an informative record of growth if one can learn to decipher them. By excavating the underground rhizomes or shoots of huckleberry plants and by identifying bud scars produced at the end of each year's growth, Thoreau was able to trace back the horizontal expansion and lateral growth of this shrub nearly a dozen years. He mused that individual huckleberry and blueberry bushes might be as old as the trees that towered above them (a conservative estimate). He also speculated that these shrubs may spread more effectively through this creeping lateral growth than by seedling establishment, which it turns out is indeed quite rare for these plants. Using a variety of ingenious ways to determine the ages of living plants and of stumps in various stages of decay, Thoreau estimated that he could trace the history of forests back through four successions or generations of trees, easily reconstructing woodlot history and dynamics back through more than a century.

To complement this history gleaned directly from the forest, Thoreau turned occasionally to written records, most notably those of William Wood, who published his observations on the landscape and native inhabitants of the New World upon his return to England from Plymouth in 1633. Thoreau located descriptions by Wood that resonated with Thoreau's own image of wild landscapes and provided a strong contrast to the tamed forests of nineteenth-century Massachusetts. These images included tall, grand, limbless trees; forests with very open understories kept clear by fires purposely set by Indians; an abundance of animals that were long gone from Thoreau's Concord, such as cougar, bear, moose, raccoon, deer, porcupine, beaver, lynx (ounce), marten, and wolf; and other animals that were greatly diminished in abundance such as squirrels, eagles, pigeons, turkeys, owls, swans, salmon, shad, and bass. Thoreau's amazement at the great abundance of animals such as geese and passenger pigeons noted in early colonial writings is matched by his keen disappointment that his own landscape was so impoverished. On the other hand, Thoreau's playful critique of Wood's rapturous enthusiasm for the New England landscape indicates that Thoreau recognized the subjective nature of all historical accounts and the proclivity of early colonists and explorers to use their letters home and writings as a means of favorable advertisement and active promotion of the New World.

For specific information on Concord, Thoreau looked at old deeds, early maps, and town records of the original "Divisions" of land that were granted to the first settlers of each town in Massachusetts; he also consulted *A History of the Town of Concord, Massachusetts,* written in 1835 by Lemuel Shattuck. In these sources he found information on the composition of the original trees that grew along the first property boundaries in town, descriptions of old land-use practices such as the development of hog pastures and woods, and confirmation of the antiquity of particular landscapes that he knew well. He delighted in his ability to confirm the age of certain place names, including "Walden," and he contrasted the current reality of many cleared fields and redeemed meadows with the primitive images suggested by place names that connote wildness.

Finally, it is very interesting to a modern ecologist who uses the tools of historical research on a regular basis to note how accurately Thoreau understood the potential for using the geological technique of stratigraphic analysis of wetland and lake sediments to interpret landscape change. Although he never developed this technique into a rigorous study, through a series of observations and explorations of wetlands he exposed many aspects of their development through the stratigraphy of materials underlying them. Toward the bottom of the ditches that the farmers and Irish laborers dug to drain swamps, Thoreau observed the remains of plants that were strikingly different from those that were growing on the surface. Such changes in plant composition with depth and the type of peat or mud beneath a swamp provided him with concrete evidence for vegetation and landscape change over many hundreds to thousands of years. Thoreau recognized that he was going back in geological time when he probed beneath a swamp, although he had few clues to the actual age of the material he discovered. The presence of "flags" (Thoreau's term for cattails) beneath a shrub or forested swamp indicated to him that meadow or open-water conditions had prevailed on the site much earlier in time. Similarly, mud underlying the peat of many wetlands suggested that many of the Concord swamps had originated as lakes that had subsequently filled in. Through time the proportion of the landscape that consisted of open ponds and lakes had diminished as this process progressively expanded the area of wetland. Layers of conifer wood retrieved from below a swamp suggested they had originated in a late-glacial boreal forest and provided concrete documentation for major changes in environment and vegetation. Although the time scales and the extent of change detected through these and other discoveries were enormous, Thoreau seemed to take the magnitude of nature's dynamics easily in stride.

In his examination of lakes and swamps Thoreau made another discovery which would later (and quite independently) be combined with the stratigraphic approach to enable scientists to resolve more precisely the major changes that have occurred in New England landscapes since the last Ice Age. Thoreau noted and actively investigated the propensity for plant pollen to be widely distributed and to accumulate on the surfaces of lakes, ponds, woodland pools, and every puddle across the land. In order to detect the presence of pine in a landscape, he stated, it was not necessary to search every woodland for trees; rather, one only needed to look at the shore of any pond in June for the distinctive yellow scum of pine pollen. Pollen of wind-pollinated plants such as pine was produced in such abundance and distributed from treetops so effectively that it reached every corner of a town and was collected on every water surface, where it floated to shore. In order to test this developing hypothesis, Thoreau repeatedly searched out small ponds and hollows quite distant from any pine and examined their shorelines carefully; in every case he discovered the pollen evidence that he was seeking. He called ponds "pollenometers," and the pollen, he noted, accumulated most abundantly along the leeward shore and in deep bays where it was driven by the wind and waves. From observations like these, and from the remarkable tendency of pollen to resist decay and to remain preserved as fossils in the deep muds and peats below lakes and swamps, comes our modern ability to reconstruct vegetation over centuries and millennia. By sampling downward in the peats from the surface of a wetland, we are going back in time, and the pollen and other fossils that we uncover provide a picture of the vegetation that grew there. The pollen of the pine and ragweed and the presence of the grass, wormwood, pigweed, and sorrel that Thoreau noted in his walks can be recovered today beneath 150 years of mud and peat in swamps and lakes. This evidence allows us to study vegetation of Thoreau's own landscape and to compare it directly to ours.

Overall, Thoreau's highly successful attempts to unravel the history of his forests and landscape provide a remarkable precursor to the reconstructive studies carried out by modern ecologists and conservationists. As he notes, every tree, every swamp, and every forest has a history. In order to observe the consequences of the varied processes that unfold in natural ecosystems over time, to interpret changes that have occurred as a result of distant events, and to understand the development of the modern landscape that surrounds us, it is necessary both to look back in time and to recognize the continuity between past and present. Historical perspectives help to explain the present; they may expose rare or uncommon processes; and they enable us to identify the important and defining events among the jumble of past occurrences. If pursued with insight, these studies provide lessons not just about nature but about human impacts on nature, lessons that may help us appreciate, manage, and conserve our landscape.

The journal entries that follow, which trace some of Thoreau's historical reconstructions, are grouped according to broad themes.

INDIANS AND ARTIFACTS

All around Walden, both in the thickest wood and where the wood has been cut off, there can be traced a meandering narrow shelf on the steep hillside, the footpath worn by the feet of Indian hunters, and still occasionally trodden by the white men, probably as old as the race of man here. And the same trail may be found encircling all our ponds.

NOVEMBER 9, 1852

Some time or other, you would say, it had rained arrowheads, for they lie all over the surface of America. You may have your peculiar tastes. Certain localities in your town may seem from association unattractive and unhabitable to you. You may wonder that the land bears any money value there, and pity some poor fellow who is said to survive in that neighborhood. But plow up a new field there, and you will find the omnipresent arrow-points strewn over it, and it will appear that the red man, with other tastes and associations, lived there too.

MARCH 28, 1859

According to Wood's "New England's Prospect," the first settlers of Concord for meat bought "venison or rockoons [raccoons]" of the Indians. The latter must have been common then.

FEBRUARY 16, 1857

I find a very regular elliptical rolled stone in the freshly (last fall) plowed low ground there, evidently brought from some pond or seaside. It is about seven inches long. The Indians prized such a stone, and I have found many of them where they haunted. Commonly one or both ends will be worn, showing that they have used it as a pestle or hammer.

MARCH 27, 1857

Some have spoken slightingly of the Indians, as a race possessing so little skill and wit, so low in the scale of humanity, and so brutish that they hardly deserved to be remembered,—using only the terms "miserable," "wretched," "pitiful," and the like. In writing their histories of this country they have so hastily disposed of this refuse of humanity (as they might have called it) which littered and defiled the shore and the interior. But even the indigenous animals are inexhaustibly interesting to us. How much more, then, the indigenous man of America! If wild men, so much more like ourselves than they are unlike, have inhabited these shores before us, we wish to know particularly what manner of men they were, how they lived here, their relation to nature, their arts and their customs, their fancies and superstitions. They paddled over these waters, they wandered in these woods, and they had their fancies and beliefs connected with the sea and the forest, which concern us quite as much as the fables of Oriental nations do. It frequently happens that the historian, though he professes more humanity than the trapper, mountain man, or gold-digger, who shoots one as a wild beast, really exhibits and practices a similar inhumanity to him, wielding a pen instead of a rifle.

FEBRUARY 3, 1859

On the same bare sand is revealed a new crop of arrowheads. I pick up two perfect ones of quartz, sharp as if just from the hands of the maker.

JANUARY 7, 1855

Returning over Great Fields, found half a dozen arrowheads, one with three scallops in the base.

MARCH 2, 1855

. . . I often think at night with inexpressible satisfaction and yearning of the *arrowheadiferous* sands of Concord. I have often spent whole afternoons, especially in the spring, pacing back and forth over a sandy field, looking for these relics of a race. This is the gold which our sands yield. The soil of that rocky spot on Simon Brown's land is quite ash-colored—now that the sod is turned up—by Indian fires, with numerous pieces of [char]coal in it. There is a great deal of this ash-colored soil in the country. We do literally plow up the hearths of a people and plant in their ashes. The ashes of their fires color much of our soil.

MAY 2, 1859

But where did the pitch pines stand originally? Who cleared the land for its seedlings to spring up in? . . . Who knows but the fires or clearings of the Indians may have to do with the presence of these trees there? They regularly cleared extensive tracts for cultivation, and these were always level tracts where the soil was light—such as they could turn over with their rude hoes. Such was the land which they are known to have cultivated extensively in this town, as the Great Fields and the rear of Mr. Dennis's,—sandy plains. It is in such places chiefly that you find their relics in any part of the county. They did not cultivate such soil as our maple swamps occupy, or such a succession of hills and dales as this oak wood covers. Other trees will grow where the pitch pine does, but the former will maintain its ground there the best. I know of no tree so likely to spread rapidly over such areas when abandoned by the aborigines as the pitch pines—and next birches and white pines.

NOVEMBER 26, 1860

A curious incident happened some four or six weeks ago which I think it worth the while to record. John [Thoreau's brother] and I had been searching for Indian relics, and been successful enough to find two arrowheads and a pestle, when, of a Sunday evening, with our heads full of the past and its remains, we strolled to the mouth of Swamp Bridge Brook. As we neared the brow of the hill forming the bank of the river, inspired by my theme, I broke forth into an extravagant eulogy on those savage times, using most violent gesticulations by way of illustration. "There on Nawshawtuck," said I, "was their lodge, the rendezvous of the tribe, and yonder, on Clamshell Hill, their feasting ground. This was, no doubt, a favorite haunt; here on this brow was an eligible lookout post. How often have they stood on this very spot, at this very hour, when the sun was sinking behind yonder woods and gilding with his last rays the waters of the Musketaquid, and pondered the day's success and the morrow's prospects, or communed with the spirit of their fathers gone before them to the land of shades!

"Here," I exclaimed, "stood Tahatawan; and there" (to complete the period) "is Tahatawan's arrowhead."

We instantly proceeded to sit down on the spot I had pointed to, and I, to carry out the joke, to lay bare an ordinary stone which my whim had selected, when lo! the first I laid hands on, the grubbing stone that was to be, proved a most perfect arrowhead, as sharp as if just from the hands of the Indian fabricator!!!

OCTOBER 29, 1837

Surveying for Sam. Pierce. Found piece of an Indian soapstone pot.

JUNE 7, 1852

Found an Indian adze in the bridle-road at the brook just beyond Daniel Clark, Jr.'s house.

JANUARY 31, 1853

J. Hosmer showed me a pestle which his son had found this summer while plowing on the plain between his house and the river. It has a rude bird's head, a hawk's or eagle's, the beak and eyes (the latter a mere prominence) serving for a knob or handle. It is affecting, as a work of art by a people who have left so few traces of themselves, a step beyond the common arrowhead and pestle and axe.

NOVEMBER 29, 1853

We survive, in one sense, in our posterity and in the continuance of our race, but when a race of men, of Indians for instance, becomes extinct, is not that the end of the world for them? Is not the world forever beginning and coming to an end, both to men and races? Suppose we were to foresee that the Saxon race to which we belong would become extinct the present winter,—disappear from the face of the earth,—would it not look to us like the end, the dissolution of the world? Such is the prospect of the Indians.

DECEMBER 29, 1853

HISTORIES AND HUMAN RECORDS

I am [reading] William Wood's "New England's Prospect." He left New England August 15th, 1633, and the last English edition referred to in this American one of 1764 is that of London, 1639.

. . . He says, "The timber of the country grows strait, and tall, some trees being twenty, some thirty foot high, before they spread forth their branches; generally the trees be not very thick, tho' there be many that will serve for mill-posts, some being three foot and an half over." One would judge from accounts that the woods were clearer than the primitive wood that is left, on account of Indian fires, for he says you might ride a-hunting in most places. "There is no underwood, saving in swamps," which the Indian fires did not burn. (*vide* Indian book.) "Here no doubt might be good done with saw mills; for I have seene of these stately high grown trees [he is speaking of pines particularly] ten miles together close by the river [probably Charles River] side." He says at first "fir [hemlock] and pine," as if the fir once grew in this part of the State abundantly, as now in Maine and further west. Of the oaks he says, "These trees afford much mast for hogs, especially every third year." Does not this imply many more of them than now? "The hornbound tree is a tough kind of wood, that requires so much pains in riving as is almost incredible, being the best to make bowls and dishes, not being subject to crack or leak," and [he] speaks both in prose and verse, of the vines being particularly inclined to run over this tree. If this is the true hornbeam it was probably larger then, but I am inclined to think it the tupelo, and that it was both larger and more abundant than commonly now, for he says it was good for bowls, and it has been so used since . . .

Of quadrupeds no longer found in Concord, he names the lion,—that Cape Ann Lion "which some affirm that they have seen," which may have been a cougar, for he adds, "Plymouth men have traded for Lions skins in former times,"—bear, moose, deer, porcupines, "the grim-fac'd Ounce, and rav'nous howling Wolf," and beaver. Martens.

"For Bears they be common, being a black kind of Bear, which be most fierce in strawberry time, at which time they have young ones; at which time likewise they will go upright like a man, and climb trees, and swim to the islands;" etc. (*vide* Indian book) In the winter they lie in "the clifts of rocks and thick swamps." The wolves hunt these in packs and "tear him as a Dog will tear a Kid." "They never prey upon English cattle, or offer to assault the person of any man," unless shot. Their meat "esteemed . . . above venison."

Complains of the wolf as the great devourer of bear, moose, and deer which kept them from multiplying more. "Of these Deer [i.e., the small] there be a great many, and more in the Massachusetts-Bay, than in any other place." "Some have killed sixteen Deer in a day upon this island," so called because the deer swam thither to avoid the wolves.

Gray squirrels were evidently more numerous than now.

I do not know whether his ounce or wild cat is the Canada lynx or wolverine. He calls it wild cat and does not describe the little wildcat. (*vide* Indian book) Says they are accounted "very good meat. Their skins be a very deep kind of fur, spotted white and black on the belly." Audubon and Bachman make the *Lynx rufus* black and white beneath. For wolf *vide* Indian book. He says: "these be killed daily in some places or other . . . Yet is there little hope of their utter destruction." "Travelling in the swamp by kennels."

Says the beaver are so cunning the English "seldom or never kill any of them, being not patient to lay a long siege" and not having experience.

Eagles are probably less common; pigeons of course (*vide* Indian book); heath cocks all gone (price "four pence"); and turkeys (good cock, "four shillings"). Probably more owls then, and cormorants, etc., etc., sea-fowl generally (of humilities he "killed twelve score at two shots"), and swans. Of pigeons, "Many of them build among the pine trees, thirty miles to the north-east of our plantations; joining nest to nest, and tree to tree by their nests, so that the Sun never sees the ground in that place, from whence the Indians fetch whole loads of them." And then for turkeys, tracking them in winter, or shooting them on their roosts at night. Of the crane, "almost as tall as a man," probably blue heron,—possibly the whooping crane or else the sandhill,—he says, "I have seen many of these fowls, yet did I never see one that was fat, though very sleaky;" neither did I. "There be likewise many Swans, which frequent the fresh ponds and rivers, seldom consorting themselves with ducks and geese; these be very good meat, the price of one is six shillings." Think of that! They had not only brant and common gray wild geese, but "a white Goose," probably the snow goose; "sometimes there will be two or three thousand in a flock;" continue six weeks after Michaelmas and return again north in March. Peabody says of the snow goose, "They are occasionally seen in Massachusetts Bay."

Sturgeon were taken at Cape Cod and in the Merrimack especially "pickled and brought to England, some of these be 12, 14, and 18 feet long." An abundance of salmon, shad, and bass,—

> The stately Bass old Neptune's fleeting post,
> That tides it out and in from sea to coast;"

"one of the best fish in the country," taken "sometimes two or three thousand at a set," "some four foot long," left on the sand behind the seine; sometimes used for manure. "Alewives . . . in the latter end of April come up to the fresh rivers to spawn, in such multitudes as is almost incredible, pressing up in such shallow waters as will scarce permit them to swim, having likewise such longing desire after the fresh water ponds, that no beatings with poles, or forcive agitations by other devices, will cause them to return to the sea, till they have cast their spawn."

"The Oysters be great ones in form of a shoe-horn some be a foot long; these breed on certain banks that are bare every spring tide. This fish without the shell is so big, that it must admit of a division before you can well get it into your mouth." For lobsters, "their plenty makes them little esteemed and seldom eaten." Speaks of "a great oyster bank" in the middle of Back Bay, just off the true mouth of the Charles, and of another in the Mistick. These obstructed the navigation of both rivers. *Vide* book for facts.

JANUARY 24, 1855

When I speak of the otter to our oldest village doctor, who should be *ex officio* our naturalist, he is greatly surprised, not knowing that such an animal is found in these parts, and I have to remind him that the Pilgrims sent home many otter skins in the first vessels that returned, together with beaver, mink, and black fox skins, and 1156 pounds of otter skins in the years 1631–36, which brought fourteen or fifteen shillings a pound, also 12,530 pounds of beaver skin. *Vide* Bradford's History of Plimouth Plantation.

DECEMBER 6, 1856

The wilderness, in the eyes of our forefathers, was a vast and howling place or *space,* where a man might roam naked of house and most other defense, exposed to wild beasts and wilder men.

MAY 5, 1859

Two hundred years ago is about as great an antiquity as we can comprehend or often have to deal with. It is nearly as good as two thousand to our imaginations. It carries us back to the days of aborigines and the Pilgrims; beyond the limits of oral testimony, to history which begins already to be enamelled with a gloss of fable, and we do not quite believe what we read; to a strange style of writing and spelling and of expression; to those ancestors whose names we do not know, and to whom we are related only as we are to the race generally. It is the age of our very oldest houses and cultivated trees.

DECEMBER 8, 1859

Looked over the oldest town records at the clerk's office this evening, the old book containing grants of land. Am surprised to find such names as "Walden Pond" and "Fair Haven" as early as 1653, and apparently 1652; also, under the first date at least, "Second Division," the rivers as North and South Rivers (no Assabet at that date), "Swamp bridge," apparently on back road, "Goose Pond," "Mr. Flints Pond," "Nutt Meadow," "Willow Swamp," "Spruce Swamp," etc., etc. "Dongy," "Dung Hole," or what-not, appears to be between Walden and Fair Haven . . . It is pleasing to read these evergreen wilderness names, *i.e.* of particular swamps and woods, then applied to now perchance cleared fields and meadows said to be redeemed.

JUNE 4, 1853

I have an old account-book, found in Deacon R. Brown's garret since his death. The first leaf or two is gone. Its cover is brown paper, on which, amid many marks and scribblings, I find written:—

"Mr. Ephraim Jones
His Wast Book
Anno Domini 1742"

It extends from November 8th, 1742, to June 20th, 1743 (inclusive) . . .

On the whole, it is remarkable how little provision was sold at the store. The inhabitants raised almost everything for themselves.

Chocolate is sold once. Rum, salt, molasses, fish, a biscuit with their drink, a little spice, and the like are all that commonly come under this head that I remember.

No butter, nor rice, nor oil, nor candles are sold. They must have used candles [of their own making], made their own butter, and done without rice. There is no more authentic history of those days than this "Wast Book" contains . . . Each line contains and states explicitly a fact. It is the best of evidence of several facts. It tells distinctly and authoritatively who sold, who bought, the article, amount, and value, and the date. You could not easily crowd more facts into one line.

JANUARY 27, 1854

Johnson in his "Wonder-working Providence" speaks of "an army of caterpillars" in New England in 1649, so great "that the cart wheels in their passage were painted green with running over the great swarms of them."

APRIL 15, 1854

WILDWOODS AND ANCIENT LANDSCAPES

That is a glorious swamp of Miles's,—the more open parts, where the dwarf andromeda [bog rosemary] prevails . . . These are the wildest and richest gardens that we have. Such a depth of verdure into which you sink. They were never cultivated by any. Descending wooded hills, you come suddenly to this beautifully level pasture, comparatively open, with a close border of high blueberry bushes. You cannot believe that this can possibly abut on any cultivated field. Some wood or pasture, at least, must intervene. Here is a place, at last, which no woodchopper nor farmer frequents and to which no cows stray, perfectly wild, where the bittern and the hawk are undisturbed. The men, women, and children who perchance come hither blueberrying in their season get more than the value of the berries in the influences of the scene.

AUGUST 5, 1852

I stand in Ebby Hubbard's yellow birch swamp, admiring some gnarled and shaggy picturesque old birches there, which send out large, knee-like limbs near the ground, while the brook, raised by the late rain, winds fuller than usual through the rocky swamp. I thought with regret how soon these trees, like the black birches that grew on the hill near by, would be all cut off, and there would be almost nothing of the old Concord left, and we should be reduced to read old deeds in order to be reminded of such things,—deeds, at least, in which some old and revered bound trees are mentioned. These will be the only proof at last that they have ever existed. Pray, farmers, keep some old woods to match the old deeds. Keep them for history's sake, as specimens of what the township was. Let us not be reduced to a mere paper evidence, to deeds kept in a chest or secretary, when not so much as the bark of the paper birch will be left for evidence, about its decayed stump.

NOVEMBER 8, 1858

He [Anthony Wright] tells me of a noted large and so-called primitive wood, Inches Wood, between the Harvard turnpike and Stow, sometimes called Stow Woods, in Boxboro and Stow.

OCTOBER 23, 1860

To Inches' Woods in Boxboro . . .

. . . there *may* be a thousand acres of old oak wood. The large wood is chiefly oak, and that white oak, though black, red, and scarlet oak are also common. White pine is in considerable quantity, and large pitch pine is scattered here and there, and saw some chestnut at the south end. Saw no hemlock or birch to speak of.

NOVEMBER 9, 1860

How many have ever heard of the Boxboro oak woods [i.e., Inches Wood]? How many have ever explored them? I have lived so long in this neighborhood and but just heard of this noble forest,—probably as fine an oak wood as there is in New England, only eight miles west of me.

A peculiarity of this, as compared with much younger woods, is that there is little or no underwood and you walk freely in every direction, though in the midst of a dense wood. You walk, in fact, *under* the wood.

Seeing this, I can realize how this country appeared when it was discovered. Such were the oak woods which the Indian threaded hereabouts.

Such a wood must have a peculiar fauna to some extent. Warblers must at least pass through it in the spring, which we do not see here.

We have but a faint conception of a full-grown oak forest stretching uninterrupted for miles, consisting of sturdy trees from one to three and even four feet in diameter, whose interlacing branches form a complete and uninterrupted canopy. Many trunks old and hollow, in which wild beasts den. Hawks nesting in the dense tops, and deer glancing between the trunks, and occasionally the Indian with a face the color of the faded oak leaf.

NOVEMBER 10, 1860

It is evident that in a wood that has been let alone for the longest period the greatest regularity and harmony in the disposition of the trees will be observed, while in our ordinary woods man has often interfered and favored the growth of other kinds than are best fitted to grow there naturally. To some, which he does not want, he allows no place at all.

NOVEMBER 16, 1860

Is it not possible that Loudon is right as it respects the primitive distribution of the birch? Are not the dense patches always such as have sprung up in open land (commonly old fields cleared by man), as is the case with the pitch pine? It disappears at length from a dense oak or pine wood. Perhaps originally it formed dense woods only where a space had been cleared for it by a burning, as now at the eastward.

MARCH 8, 1861

PLANT, TREE, AND FOREST AGE

Went into Tommy Wheeler's house, where still stands the spinning-wheel, and even the loom, home-made. Great pitch pine timbers overhead, fifteen or sixteen inches in diameter, telling of the primitive forest here.

DECEMBER 28, 1851

I examine that oak lot of Rice's next to the pine strip of the 16th. The oaks (at the southern end) are about a dozen years old. As I expected, I find the stumps of the pines which stood there before quite fresh and distinct, not much decayed, and I find by their rings that they were about forty years old when cut, while the pines which sprang from [them] are now about twenty-five or thirty. But further, and unexpectedly, I find the stumps, in great numbers, now much decayed, of an oak wood which stood there more than sixty years ago. They are mostly shells, the sap-wood rotted off and the inside turned to mould. Thus I distinguished four successions of trees . . .

It is easier far to recover the history of the trees which stood here a century or more ago than it is to recover the history of the men who walked beneath them. How much do we know—how little more can we know—of these two centuries of Concord life?

OCTOBER 19, 1860

I examined the huckleberry bushes next the wall in that same dense pitch and white pine strip. I found that the oldest bushes were about two feet high and some eight or ten years old, and digging with spade and hands, I found that their roots did not go deep, but that they spread by a vigorous shoot which forked several times, running just under the leaves or in the surface soil, so that they could be easily pulled up. One ran seven feet . . . or more in length. And three or four bushes stood on this shoot, and though these bushes after a few years did not grow more than an inch in a year, these subterranean shoots had grown six or twelve inches at the end, and there seemed to be all the vigor of the plant. The largest bushes preserved still a trace of their origin from a subterranean shoot, the limbs being one-sided and the brash aslant. It is very likely, then, if not certain, that these roots are as old as the pine wood which over-shadows them; or it is so long since the seedling huckleberry came up there.

OCTOBER 19, 1860

I could tell a white pine here when it was for the most part a mere rotten mound, by the regularity crosswise of the long knots a foot from the ground in the top of the rotten core, representing the peculiarly regular branches of the little white pine and the best preserved as the hardest and pitchiest part.

OCTOBER 22, 1860

. . . The only other ancient traces of trees were perhaps the semiconical mounds which had been heaved up by trees which fell in some hurricane.

OCTOBER 20, 1860

But I have not yet taken into the account the fact that, though the thickness of the layer is less, its superficies, or extent, is greater, as the diameter of the tree increases. Let us compare the three portions of wood. If the diameter at the end of the first fifty years is four, the second fifty, six, and the third fifty, seven, then the amount of wood added each term will be (to omit very minute fractions) twelve and a half, fifteen and a half, and ten respectively. So that, though in the second fifty the rings are twice as near together, yet considerably more wood is produced than in the first, but in the third fifty the tree is evidently enfeebled, and it probably is not profitable (so far as bulk is concerned) to let it grow any more.

NOVEMBER 1, 1860

Wetherbee's oak lot may contain four or five acres. The trees are white, red, scarlet, and swamp white oaks, maple, white pine, and ash. They are unusually large and old. Indeed, I doubt if there is another hereabouts of oaks as large. It is said that Wetherbee left them for the sake of mast for pigeons . . .

As you approach the wood, and even walk through it, the trees do not affect you as large, but as surely as you go quite up to one you are surprised. The very lichens and mosses which cover the rocks under these trees seem, and probably are in some respects, peculiar. Such a wood, at the same time that it suggests antiquity, imparts an unusual dignity to the earth.

NOVEMBER 2, 1860

SWAMP PEAT AND LAKE MUD

I improve the dry weather to examine the middle of Gowing's Swamp. There is in the middle an open pool, twenty or thirty feet in diameter, nearly full of sphagnum and green froth on the surface (frog-spittle), and what other plants I could not see on account of the danger in standing on the quaking ground; then a dense border, a rod or more wide, of a peculiar rush (?), with clusters of seed-vessels, three together, now going to seed, a yellow green, forming an abrupt edge next the water, this on a dense bed of quaking sphagnum, in which I sink eighteen inches in water, upheld by its matted roots, where I fear to break through. On this the spatulate sundew abounds. This is marked by the paths of muskrats, which also extend through the green froth of the pool. Next comes, half a dozen rods wide, a dense bed of *Andromeda calyculata* [leather leaf],—the *A. Polifolia* mingled with it,—the rusty cottongrass, cranberries,—the common and also *V. Oxycoccus,*—pitcher-plants, sedges, and a few young spruce and larch here and there,—all on sphagnum, which forms little hillocks about the stems of the andromeda. Then ferns, now yellowing, high blueberry bushes, etc., etc., etc.,—or the bushy and main body of the swamp, under which the sphagnum is now dry and white.

AUGUST 23, 1854

Measured Gowing's Swamp two and a half rods northeast of the middle of the hole, *i.e.* in the andromeda and sphagnum near its edge, where I stand in the summer; also five rods northeast of the middle of the open hole, or in the midst of the andromeda. In both these places the pole went hard at first, but broke through a crust of roots and sphagnum at about three feet beneath the surface, and I then easily pushed the pole down just twenty feet. This being a small pole, I could not push it any further holding it by the small end; it bent then. With a longer and stiffer pole I could probably have fathomed thirty feet. It seems, then, that there is, over this andromeda swamp, a crust about three feet thick, of sphagnum, andromeda [*Andromeda calyculata* and *Polifolia*] and *Kalmia glauca* [sheep laurel], etc., beneath which there is almost clear water, and, under that, an exceedingly thin mud. There can be no soil above that mud, and yet there were three or four larch trees three feet high or more between these holes, or over exactly the same water, and there were small spruces near by. For aught that appears, the swamp is as deep under the andromeda as in the middle. The two andromedas and the *Kalmia glauca* may be more truly said to grow in water than in soil there. When the surface of a swamp shakes for a rod around you, you may conclude that it is a network of roots two or three feet thick resting on water or a very thin mud. The surface of that swamp, composed in great part of sphagnum, . . . which floats on the surface of clear water, and, accumulating, at length affords a basis for that large-seeded sedge (?), andromedas, etc. The filling up of a swamp, then, in this case at least, is not the result of a deposition of vegetable matter washed into it, settling to the bottom and leaving the surface clear, so filling it up from the bottom to the top; but the vegetation first extends itself over it as a film, which gradually thickens till it supports shrubs and completely conceals the water, and the under part of this crust drops to the bottom, so that it is filled up first at the top and bottom, and the middle part is the last to be reclaimed from the water.

Perhaps this swamp is in the process of becoming peat. This swamp has been partially drained by a ditch. I fathomed also two rods within the edge of the blueberry bushes, in the path, but I could not force a pole down more than eight feet five inches; so it is much more solid there, and the blueberry bushes require a firmer soil than the water andromeda.

This is a regular *quag,* or shaking surface, and in this way, evidently, floating islands are formed. I am not sure but that meadow, with all its bushes in it, would float a man-of-war.

FEBRUARY 1, 1858

A man at work on the Ledum Pond, draining it, says that, when they had ditched about six feet deep, or to the bottom, near the edge of this swamp, they came to old flags [i.e., cattails], and he thought that the whole swamp was once a pond and the flags grew by the edge of it. Thought the mud was twenty feet deep near the pool, and that he had found three growths of spruce, one above another, there.

OCTOBER 23, 1858

I can sound the swamps and meadows on the line of the new road to Bedford with a pole, as if they were water. It may be hard to break through the crust, but then it costs a very slight effort to force it down, sometimes nine or ten feet, where the surface is dry. Cut a straight sapling, an inch or more in [diameter]; sharpen and peel it that it may go down with the least obstruction.

JULY 6, 1853

Rice and others are getting out mud in the pond-hole opposite Breed's. They have cut down straight through clear black muck, perfectly rotted, eight feet, and it is soft yet further. Button-bushes, andromeda, proserpinaca, hardhack, etc., etc., grow atop. It looks like a great sponge. Old trees buried in it.

AUGUST 27, 1854

POLLEN AND THE PLANTS IT REPRESENTS

There is a dust on Walden,—where I come to drink,—which I think is the pollen of such trees and shrubs as are now in blossom,—aspens, maples, sweet-fern, etc.,—food for fishes.

MAY 5, 1852

You may say that the oaks (all but the chestnut oak I have seen) were in bloom yesterday; *i.e.,* shed pollen more or less. Their blooming is soon over.

MAY 23, 1854

There is considerable pollen on the pond [Walden]; more than last year, notwithstanding that all the white pines near the pond are gone [having been cut] and there are very few pitch. It must all come from the pitch pine, whose sterile blossoms are now dry and empty, for it is earlier than the white pine. Probably I have never observed it in the river because it is carried away by the current.

JUNE 15, 1852

Went, across lots still, to Monadnock, the base some half-dozen miles in a straight line from Peterboro,—six or seven miles. (It had been eleven miles *(by road)* from Mason Village to Peterboro.) My clothes sprinkled with ambrosia [ragweed] pollen.

SEPTEMBER 7, 1852

Having noticed the pine pollen washed up on the shore of three or four ponds in the woods lately and at Ripple Lake, a dozen rods from the nearest pine, also having seen the pollen carried off visibly half a dozen rods from a pitch pine which I had jarred, and rising all the while when there was very little wind, it suggested to me that the air must be full of this fine dust at this season, that it must be carried to great distances, when dry, and falling at night perhaps, or with a change in the atmosphere, its presence might be detected remote from pines by examining the edges of pretty large bodies of water, where it would be collected to one side by the wind and waves from a large area.

So I thought over all the small ponds in the township in order to select one or more most remote from the woods or pines, whose shores I might examine and so test my theory. I could think of none more favorable than this little pond only four rods [one rod equals 16½ feet] in diameter, a watering-place in John Brown's pasture, which has but few [lily] pads in it. It is a small round pond at the bottom of a hollow in the midst of a perfectly bare, dry pasture. The nearest wood of any kind is just thirty-nine rods distant northward, and across a road from the edge of the pond. Any other wood in other directions is five or six times as far. I knew it was a bad time to try my experiment,—just after such heavy rains and when the pines are effete,—a little too late. The wind was now blowing quite strong from the northeast, whereas all the pollen that I had seen hitherto had been collected on the northeast sides of ponds by a southwest wind. I approached the pond from the northeast and, looking over it and carefully along the shore there, could detect no pollen. I then proceeded to walk round it, but still could detect none. I then said to myself, If there was any here before the rain and northeast wind, it must have been on the northeast side and then have washed over and now up high quite at or on the shore. I looked there carefully, stooping down, and was gratified to find, after all, a distinct yellow line of pollen dust about half an inch in width—or washing off to two or three times that width—quite on the edge, and some dead twigs which I took up from the wet shore were completely coated with it, as with sulphur. This yellow line reached half a rod along the southwest side, and I then detected a little of the dust slightly graying the surface for two or three feet out there. . . .

When I thought I had failed, I was much pleased to detect, after all, this distinct yellow line, revealing unmistakably the presence of pines in the neighborhood and thus confirming my theory. As chemists detect the presence of ozone in the atmosphere by exposing it to a delicately prepared paper, so the lakes detect for us thus the presence of the pine pollen in the atmosphere. They are our *pollenometers.* How much of this invisible dust must be floating in the atmosphere, and be inhaled and drunk by us at this season!! Who knows but the pollen of some plants may be unwholesome to inhale and produce the diseases of the season?

Of course a large pond will collect the most, and you will find most at the bottom of long deep bays into which the wind blows.

I do not believe that there is any part of this town on which the pollen of pine may not fall.

JUNE 21, 1860

Landscape Change

There is scarcely a wood of sufficient size and density left now for an owl to haunt in, and if I hear one hoot I may be sure where he is.

NOVEMBER 28, 1859

One of the most striking and sobering lessons that Thoreau learned from his historical reconstructions was the extent to which the land-use practices of the seventeenth, eighteenth, and nineteenth centuries had changed the natural landscape of New England. This lesson was all too apparent in his favorite woodland at Walden Pond. Although the pond and its forest landscape are perceived as a nature retreat by most readers of Thoreau, he certainly was not deceived about the ravaged history of the land surrounding his cabin or the faint resemblance that the site held to its original splendor. Gone were the lofty pines and oak that some older citizens of Concord remembered and often told about in their conversations. Gone too were the dark, dense forests, the sheltered bowers and coves that Thoreau himself could recall from his youth. In their place was a cut-over, second-growth forest that woodchoppers had laid to waste in the past and would continue to cut throughout the years of Thoreau's two-mile journeys from his home in the center of Concord to Walden's shores.

The woods at Walden formed one of Ralph Waldo Emerson's woodlots and were purchased by Emerson as a possible site for a house of his own and to afford them some protection from relentless cutting. Like all the remaining woodlands in eastern Massachusetts, the forests that Emerson and most other Concord citizens owned had been cut frequently. Consequently, the Walden woods had been substantially transformed from their condition at the time of the town's founding. The hillsides of cut stumps, sprouting oaks, and young pines were a constant reminder of human impact. Thoreau's bean field was the legacy of earlier clearing and a modest attempt at agriculture along the pond's shore. The Fitchburg Railroad (running on tracks constructed just three years before Thoreau "retreated" to Walden), with its locomotive that screamed across the far end of the pond, was another manifestation of the social and technological advancement in America that demanded more resources from the land. In all these features of his landscape, Thoreau could gauge the extent of past change and anticipate future ones.

The rate at which the landscape and human land-use practices were changing in the early nineteenth century was dramatic. For some of Thoreau's old-timer acquaintances, the world had been "turned upside down" by a lifetime of change; much of the landscape had been completely transformed from forest wilderness to cultivated land and village. By 1830 open agricultural land of pasture, hay mowings, tilled land, and wet meadows had become the predominant condition of the countryside around Concord, and the forest was restricted to the poorest of agricultural sites such as swamps, the dry sand plains around Walden, and the least accessible areas. But in even distant recesses and remote locations Thoreau could usually detect evidence that the land had been disturbed by people or their grazing animals. The rockiest hillsides bore the signs of old cart trails that were used for the annual retrieval of fuelwood or paths that had been cut by cows as they foraged through the woods and brush. The extensive freshwater marshes that bordered the sluggish rivers had been drained by ditches, browsed by wide-ranging cattle, and repeatedly cut and burned by humans to yield the productive meadowlands that characterized Concord. As Thoreau observed on his walks, the imprint of human activity was so great across New England that it seemed as if all the land was confined and bounded by stone walls, wooden fences, and ownership boundaries that had been surveyed many times over.

In a few places, vestiges of the original landscape remained. The huge pasture oaks that stood singly, in small groups, or in broad open stands were reminiscent of European deer parks and were interpreted (perhaps incorrectly) by Thoreau to be remnants of forests that originally covered the land more densely. Many forested swamps, though cut over, were fairly unaltered and retained a sense of wildness that Thoreau identified with an earlier state. Elsewhere, the remaining "natural landscape" was either a faint second-growth shadow of the original forest or was utterly and completely modified. Although Thoreau found the "new" white pine woods that developed on abandoned fields across Concord to be beautiful, he acknowledged that the shape, growth, and size of the individual trees were greatly diminished in comparison with the huge white pine logs that had traveled through the town center on the rail line from northern New England and New York.

As an example of the enduring impact that human land use exerted on nature, Thoreau recorded in his journal the history of the treeless summit of Mount Monadnock in New Hampshire. This isolated mountain forms a striking conical bulge on the horizon of any northern view from eastern Massachusetts, and it continues to attract many hikers to its popular summit, as it did Thoreau on a number of occasions. According to the local story that he was told on one trip, in the eighteenth century fires were purposely set on the mountain by farmers to consume the original forest and brush in an attempt to eliminate the protective cover for wolves that were threatening livestock in the surrounding valleys. The wolves were successfully driven off by this destruction, but the trees have been extraordinarily slow to reestablish across the bare slopes in Monadnock's harsh and exposed environment. As a result, the farmers inadvertently created a permanently bald mountain peak. The duration of this major change in the vegetation of one of New England's most prominent and popular summits was an indication to Thoreau of the enormous scale and long-

lasting legacy of human actions, even those that are indirect. He realized that, because many years had passed and the actual evidence of the fires had faded away, it would be possible to mistake the treeless summit as a natural condition, whereas it actually had been created by human action. This observation by Thoreau contains a powerful message concerning the interpretation of the modern American landscape. Despite a strongly "natural" appearance, most of our land and waters have been shaped directly or indirectly into their current condition by cultural forces.

The transformation in Thoreau's landscape was perhaps most striking in its effect on animals. Thoreau laments, in some of the strongest terms that appear in his writings, the absence of the noble, larger animals—the lynx, bear, beaver, moose, deer, and wolf—and the ease with which humankind accepts this tamed and imperfect condition as natural. Interestingly, he looked beyond the loss of the greater furbearing animals to comment on changes that had occurred in the lesser species as well. In Minott's lifetime, he reported, the squirrel population had apparently declined to one-tenth its original size. Thoreau also speculated that along with the large mammals even such small insects as the black fly and the "no-see-em," which he encountered in abundance on his trips to the wilds of Maine, had also disappeared from southern New England as a consequence of the spread of "civilization." Given the range and intensity of his interest, one can only wonder whether Thoreau would be gratified to learn that both of these insect pests have accompanied the bear, moose, deer, and eagle by the multitude in their reappearance in the New England landscape as the forests have regrown to cover the countryside.

Thoreau provides us with important insights into the making of a cultural landscape when he addresses the legacy of his colonial predecessors in his own countryside. When Thoreau rather offhandedly describes long-abandoned practices such as allowing hogs to run wild in the woods, the selling of tanbark in the streets, and the burning of forests to eliminate wolves, he is registering his recognition that many of the important and long-lasting human activities that had altered nature were no longer pursued during his own lifetime. Thus Thoreau was forced to use his own imagination, in conjunction with historical research and insight, in order to interpret the present.

In his lengthy description of the felling of a large white pine that he viewed one day from the cliff on Fair Haven Hill, Thoreau provides an evocative and suspenseful eulogy to nature as well as a striking testimony of change. The four-foot-diameter tree was one of only a dozen such noble pines that Thoreau knew to be left in Concord, and its girth was a challenge to the two woodchoppers with their cross-cut saw whom

Thoreau secretly observed from above. From his distant vantage point, through the lengthy process of felling the tree, Thoreau romanticized the life and antiquity of the pine, which he estimated to be two centuries old. As the men toiled, he sang the tree's praises, invoking its stature and majesty. With it, he wrote, fell some of Concord's original wildness: a loss that shook history as the tree shook the earth when it fell. Thoreau's extended description of this process is one of his most gripping tales of humans interacting with nature. And yet, in the end, Thoreau the scientist put the actual history of the tree into perspective when he finally clambered down the slope to view the pine more closely and to age it precisely. When he counted the number of growth rings from the center to the outer edge, he determined that the 48 inches of diameter growth on this white pine was produced in a mere 94 years! Remarkably, these giant trees dated only to the late 1750s, some 115 years after Concord was established. In an instant some of Thoreau's romance was cut short, for these trees were not ancient remnants of a lost wilderness but rather a product of forest succession and commonplace processes that Thoreau knew well. Through this bit of sleuthing the natural historian of Concord learned first-hand just how rapidly the land could change and how deceptive this change could be.

> Hardly a rood of land but can show its fresh wound or indelible scar, in proof that earlier or later man has been there.
>
> MARCH 14, 1838

> Looking across the river to Conantum from the open plains, I think how the history of the hills would read, since they have been pastured by cows, if every plowing and mowing and sowing and chopping were recorded.
>
> JULY 13, 1851

> On the most retired, the wildest and craggiest, most precipitous hillside you will find some old road by which the teamster carted off the wood.
>
> APRIL 22, 1852

> What are our fields but *felds* or *felled* woods. They bear a more recent name than the woods, suggesting that previously the earth was covered with woods. Always in the new country a field is a clearing.
>
> JANUARY 27, 1853

A man in Peterboro told me that his father told him that [Mount] Monadnock used to be covered with forest, that fires ran through it and killed the turf; then the trees were blown down, and their roots turned up and formed a dense and impenetrable thicket in which the wolves abounded. They came down at night, killed sheep, etc., and returned to their dens, whither they could not be pursued, before morning; till finally they set fire to this thicket, and it made the greatest fire they had ever had in the county, and drove out all the wolves, which have not troubled them since.

SEPTEMBER 6, 1852

In these respects those Maine woods differed essentially from ours. There you are never reminded that the wilderness which you are threading is, after all, some villager's familiar wood-lot from which his ancestors have sledded their fuel for generations, or some widow's thirds, minutely described in some old deed, which is recorded, of which the owner has got a plan, too, and old bound marks may be found every forty rods if you will search. What a history this Concord wilderness which I affect so much may have had! How many old deeds describe it,—some particular wild spot,—how it passed from Cole to Robinson, and Robinson to Jones, and Jones finally to Smith, in course of years! Some have cut it over three times during their lives, and some burned it and sowed it with rye, and built walls and made a pasture of it, perchance. All have renewed the bounds and reblazed the trees many times.

JANUARY 1, 1858

In the wildwood at least there are commonly only fires and insects or blight, and not the axe and plow and the cattle, to interrupt the regular progress of things.

OCTOBER 16, 1860

Swamps are, of course, least changed with us,—are nearest to their primitive state of any woodland. Commonly they have only been cut, not redeemed.

OCTOBER 22, 1860

On the sandy slope of the cut, close by the pond, I notice the chips which some Indian fletcher [arrow maker] has made. Yet our poets and philosophers regret that we have no antiquities in America, no ruins to remind us of the past. Hardly can the wind blow away the surface anywhere, exposing the spotless sand, even though the thickest woods have recently stood there, but these little stone chips made by some aboriginal fletcher are revealed.

OCTOBER 15, 1858

An old man who used to frequent Walden fifty-five years ago, when it was dark with surrounding forests, tells me that in those days he sometimes saw it all alive with ducks and other game. He went there to fish and used an old log canoe, made of two white pine logs dug out and pinned together and pitched, which he found on the shore. It was very clumsy but durable and belonged to the pond. He did not know whom it belonged to; it belonged to the pond.

JUNE 16, 1853

I loved to hear of the old log canoe, which perchance had first been a tree on its brink, and then, as it were, fell into the water, to float there for a generation as the only proper vessel for it,—very thick and at length water-logged. So primitive a vessel! I remember that when I first paddled on it there were more large trunks of trees to be seen indistinctly lying on the bottom, which had probably blown over formerly, when the trees were larger, or had been left on the ice at the last cutting, when wood was cheaper; but now for the most part they have disappeared. The old log canoe, which took the place of a more graceful one of Indian construction.

Now the trunks of trees on the bottom and the old log canoe are gone, the dark surrounding woods are gone, and the villagers, who scarcely know how it lies, instead of going to the pond to bathe or drink, are thinking to bring its water to the village in a pipe, to form a reservoir as high as the roofs of the houses, to wash their dishes and be their scullion [dishwasher],—which should be more sacred than the Ganges,—to earn their Walden by the turning of a cock or drawing of a plug, as they draw cider from a cask.

JUNE 17, 1853

Wyman told Minott that . . . One day he saw a bull on the northerly side [of Walden] swim across to get at some cows on the south.

MAY 4, 1857

He [Minott] remembers when a Prescott, who lived where E. Hosmer does, used to let his hogs run in the woods in the fall, and they grew quite fat on the acorns, etc., they found, but now there are few nuts, and it is against the law . . . And yet the gray squirrels were ten then to one now. But for the most part, he says, the world is turned upside down.

OCTOBER 12, 1851

It is observable that not only the moose and the wolf disappear before the civilized man, but even many species of insects, such as the black fly and the almost microscopic "no-see-em." How imperfect a notion have we commonly of what was the actual condition of the place where we dwell, three centuries ago!

JANUARY 29, 1856

Do not the forest and the meadow now lack expression, now that I never see nor think of the moose with a lesser forest on his head in the one, nor of the beaver in the other? When I think what were the various sounds and notes, the migrations and works, and changes of fur and plumage which ushered in the spring and marked the other seasons of the year, I am reminded that this my life in nature, this particular round of natural phenomena which I call a year, is lamentably incomplete. I listen to [a] concert in which so many parts are wanting. The whole civilized country is to some extent turned into a city, and I am that citizen whom I pity. Many of those animal migrations and other phenomena by which the Indians marked the season are no longer to be observed. I seek acquaintance with Nature,—to know her moods and manners. Primitive Nature is the most interesting to me. I take infinite pains to know all the phenomena of the spring, for instance, thinking that I have here the entire poem, and then, to my chagrin, I hear that it is but an imperfect copy that I possess and have read, that my ancestors have torn out many of the first leaves and grandest passages, and mutilated it in many places. I should not like to think that some demigod had come before me and picked out some of the best of the stars. I wish to know an entire heaven and an entire earth. All the great trees and beasts, fishes and fowl are gone. The streams, perchance, are somewhat shrunk.

MARCH 23, 1856

There are some very handsome white pines and pine groves on the left of the road just before you enter the woods. They are of second growth, of course, broad and perfect, with limbs almost to the ground, and almost as broad as they are high, their fine leaves trembling with silvery light, very different from the tall masts of the primitive wood, naked of limbs beneath and crowded together.

JUNE 23, 1852

I was just roused from my writing by the [railroad] engine's whistle, and, looking out, saw shooting through the town two enormous pine sticks stripped of their bark, just from the Northwest and going to Portsmouth Navy-Yard, they say. Before I could call Sophia [Thoreau's sister], they had got round the curve and only showed their ends on their way to the Deep Cut. Not a tree grows now in Concord to compare with them. They suggest what a country we have got to back us up that way. A hundred years ago or more perchance the wind wafted a little winged seed out of its cone to some favorable spot and this is the result. In ten minutes they were through the township, and perhaps not half a dozen Concord eyes rested on them during their transit.

JUNE 23, 1853

A hundred years ago, as I learned from Ephraim Jones's ledger, they sold bark in our street. He gives credit for a load. Methinks my genius is coeval with that time. That is no great wildness or *selvaggia* that cannot furnish a load of bark, when the forest has lost its shagginess. This is an attempt to import this wildness into the cities in a thousand shapes. Bark is carried thither by ship and by cartloads. Bark contains the principle of tannin, by which not only the fibre of skins but of men's thoughts is hardened and consolidated. It was then that a voice was given to the dog, and a manly tone to the human voice. Ah! already I shudder for these comparatively degenerate days of the village, when you cannot collect a load of bark of good thickness.

JANUARY 31, 1854

This afternoon, being on Fair Haven Hill, I heard the sound of a saw, and soon after from the Cliff saw two men sawing down a noble pine beneath, about forty rods off. I resolved to watch it till it fell, the last of a dozen or more which were left when the forest was cut and for fifteen years have waved in solitary majesty over the sprout-land. I saw them like beavers or insects gnawing at the trunk of this noble tree, the diminutive manikins with their cross-cut saw which could scarcely span it. It towered up a hundred feet as I afterward found by measurement, one of the tallest probably in the township and straight as an arrow, but slanting a little toward the hillside, its top seen against the frozen river and the hills of Conantum. I watch closely to see when it begins to move. Now the sawyers stop, and with an axe open it a little on the side toward which it leans, that it may break the faster. And now their saw goes again. Now surely it is going; it is inclined one quarter of the quadrant, and, breathless, I expect its crashing fall. But no, I was mistaken; it has not moved an inch; it stands at the same angle as at first. It is fifteen minutes yet to its fall. Still its branches wave in the wind, as if it were destined to stand for a century, and the wind soughs through its needles as of yore; it is still a forest tree, the most majestic tree that waves over Musketaquid. The silvery sheen of the sunlight is reflected from its needles; it still affords an inaccessible crotch for the squirrel's nest; not a lichen has forsaken its mast-like stem, its raking mast,—the hill is the hulk. Now, now's the moment! The manikins at its base are fleeing from their crime. They have dropped the guilty saw and axe. How slowly and majestically it starts! as if it were only swayed by a summer breeze, and would return without a sigh to its location in the air. And now it fans the hillside with its fall, and it lies down to its bed in the valley, from which it is never to rise, as softly as a feather, folding its green mantle about it like a warrior, as if, tired of standing, it embraced the earth with silent joy, returning its elements to the dust again. But hark! there you only saw, but did not hear. There now comes up a deafening crash to these rocks, advertising you that even trees do not die without a groan. It rushes to embrace the earth, and mingle its elements with the dust. And now all is still once more and forever, both to eye and ear.

I went down and measured it. It was about four feet in diameter where it was sawed, about one hundred feet long. Before I had reached it the axemen had already half divested it of its branches. Its gracefully spreading top was a perfect wreck on the hillside as if it had been made of glass, and the tender cones of one year's growth upon its summit appealed in vain and too late to the mercy of the chopper. Already he has measured it with his axe, and marked off the mill-logs it will make. And the space it occupied in upper air is vacant for the next two centuries. It is lumber. He has laid waste the air. When the fish hawk in the spring revisits the banks of the Musketaquid, he will circle in vain to find his accustomed perch, and the hen-hawk will mourn for the pines lofty enough to protect her brood. A plant which it has taken two centuries to perfect, rising by slow stages into the heavens, has this afternoon ceased to exist. Its sapling top had expanded to this January thaw as the forerunner of summers to come. Why does not the village bell sound a knell? I hear no knell tolled, I see no procession of mourners in the streets, or the woodland aisles. The squirrel has leaped to another tree; the hawk has circled further off, and has now settled upon a new eyrie, but the woodman is preparing [to] lay his axe at the root of that also.

DECEMBER 30, 1851

The pine I saw fall yesterday measured to-day one hundred and five feet, and was about ninety-four years old. There was one still larger lying beside it, one hundred and fifteen feet long, ninety-six years old, four feet diameter the longest way. The tears were streaming from the sap-wood—about twenty circles—of each, pure amber or pearly tears.

DECEMBER 31, 1851

Insights into the Ecology and Conservation of the Land

When I walk in the fields of Concord and meditate on the destiny of this prosperous slip of the Saxon family, the unexhausted energies of this new country, I forget that this which is now Concord was once Musketaquid, and that the *American race* has had its destiny also. Everywhere in the fields, in the corn and grain land, the earth is strewn with the relics of a race which has vanished as completely as if trodden in with the earth. I find it good to remember the eternity behind me as well as the eternity before. Wherever I go, I tread in the tracks of the Indian. I pick up the bolt which he has but just dropped at my feet. And if I consider destiny I am on his trail. I scatter his hearthstones with my feet, and pick out of the embers of his fire the simple but enduring implements of the wigwam and the chase. In planting my corn in the same furrow which yielded its increase to his support so long, I displace some memorial of him.

MARCH 19, 1842

The steam whistle at a distance sounds even like the hum of a bee in a flower. So man's works fall into nature.

JUNE 12, 1852

What are the natural features which make a township handsome? A river, with its waterfalls and meadows, a lake, a hill, a cliff or individual rocks, a forest, and ancient trees standing singly. Such things are beautiful; they have a high use which dollars and cents never represent. If the inhabitants of a town were wise, they would seek to preserve these things, though at a considerable expense; for such things educate far more than any hired teachers or preachers, or any at present recognized system of school education. I do not think him fit to be the founder of a state or even of a town who does not foresee the use of these things, but legislates chiefly for oxen, as it were.

JANUARY 3, 1861

We can work our way through the underlying chronology of Thoreau's journals, starting from his retrospective view back in time to the Native Americans and then the first glimpses of the land by the early colonists, progressing to the agricultural heyday of New England in his own time, followed by the establishment of new forests on abandoned land, and concluding with his thoughts on the extent of change in New England forests. The series of observations and insights that Thoreau and his approach to reading nature provide for interpreters of the modern landscape are both manifold and profound.

At the most basic level, Thoreau gives us an approach to interpreting nature that is based on a combination of retrospective studies of the past, a sound understanding of the biology of organisms and physical characteristics of the environment, and an awareness of human history and activity. Thoreau applied this method daily in his own interpretation of the New England landscape, and his journals provide us with broad insights into many of the important and pervasive factors that control the patterns and variation in the natural world. Major themes that emerge in these daily descriptions include the power of time and the natural and human tendency toward change. Moreover, since one dominant lesson that emerges from Thoreau's observations is that historical factors and past land-use activity determine current conditions, we can use the detailed descriptions in his journals of people, daily activities, and the land to understand our ancestral landscape and the land-use history that have given rise to the countryside we currently inhabit. Indeed, the fact that so many important insights into natural history emerge from the writings of a nineteenth-century naturalist should encourage us to continue to peruse old books, maps, and forgotten studies to glean information that may be directly relevant to our day.

Finally, Thoreau helps us to appreciate the aesthetic value of the cultural landscape in America. It is ironic to recognize today, when a high value is placed on nature, wilderness, and old-growth landscape, that America's premier nature writer and propounder of conservation and wilderness values lived at a time when the New England landscape was arguably the most tamed and most dominated by human activity in its entire history. Every site that he saw, every wood and swamp and meadow that he walked for inspiration, was shaped by a combination of human and natural forces. Walden was a woodlot whose solitude was disturbed by the ring of the axe and the scream of the locomotive. Nevertheless, despite the overwhelming presence of people in everything that he viewed, Thoreau was able to find nature and an element of wildness everywhere he looked and to appreciate and even thrive on the merging of natural and cultural influences.

For the natural historian, Thoreau offers an approach to interpreting the various factors that control natural patterns, including the distribution of specific plants, the arrangement of vegetation and animals in the landscape, and the changes that have occurred in nature through time. From Thoreau's perspective, the natural history of a landscape involves the interplay of three factors: the inherent characteristics of species and the land, the history of previous impact by people and natural processes, and time. Each tree, each forest, and every landscape has a history, and the only way to truly understand their modern condition is to evaluate their past.

A major theme in Thoreau's landscape was change. He described and lamented the forces that had turned an Indian wilderness into the tamed land of New England, and he and his companions could also see their own landscape being transformed before them. A modest amount of this change was the consequence of natural processes such as insect outbreaks or windstorms, but the major force was clearly human. In nineteenth-century New England, the forests changed virtually overnight as the direct consequence of cutting, burning, clearing, and grazing. Even when left alone, they continued to change through a process of growth and succession as they gradually recovered from a host of prior human impacts. Thus, through his own experience and observations, Thoreau relates an important message: we must expect change in nature, and much of this will be unanticipated and will relate to earlier processes that no longer operate in the landscape and that we can do nothing to control.

This message has at least two cautionary notes for the practicing or recreational conservationist. First, it must be recognized that if we set out with the expectation of protecting and preserving any landscape as it is today, we are certain to be frustrated, for it will inevitably continue to change. Second, if we seek to recreate or maintain "natural" features of our landscape, we need to understand fully the processes that have formed them in the past and then be willing either to replicate them or to develop equivalent substitutes for them. Perhaps the best examples of these lessons are the commonplace forests of pine that Thoreau observed in his studies of succession and the varied grasslands that he wandered through on his daily walks. Woods of white pine and pitch pine have been a delightful and characteristic part of the New England landscape since Thoreau's day, but they have become less common in the twentieth century. If we set out to preserve these forests as reserves and as a "natural" part of the New England landscape, we will be disappointed with the long-term result unless we are willing to recognize the inevitable developments that will take place within them. As Thoreau so clearly demonstrated, these pine forests are largely a historical accident of human history and plant biology; they proliferated in the New England landscape as a result of cultural activities, especially the broad-scale abandonment of old pastures and fields. Thus, from Cape Cod to Thoreau's Concord to the northern hills of Vermont and New Hampshire, we may note, as Thoreau did, that beneath the overstory of tall pine is a new, growing forest of oaks, maples, birch, and hemlock that is waiting to take over when the pines fall from age, wind, or the saw.

This message was brought home vividly to New England in 1985 when a tornado struck western Connecticut and destroyed a majestic pine forest. The Cathedral Pines, located in the village of Cornwall and owned by The Nature Conservancy, was one of the most ancient old-field white pine forests in New England, composed of trees 150 to 300 years in age and more than three feet in diameter. Although the impression was misleading, this forest, with its tall, dark understory, conveyed all the splendor of an old-growth, natural forest. When blown down, the 20 acres of pine were replaced, through natural regeneration and growth, by a new forest consisting of hardwoods and hemlock but almost completely lacking in white pine. To recreate such a majestic pine stand, we would need to restore the agricultural practices that gave rise to it or devise alternative management regimes. The same would be true if we sought to recreate the bucolic scenes

from Thoreau's day of scattered fencerow oaks and hickories with their widespread branches, openly grazed forest, or coppiced chestnut sprouts. In the modern landscape, where farmers and cows are increasingly scarce, locating the knowledge and resources to undertake such tasks is a formidable challenge.

A parallel situation exists for the grasslands, meadows, and shrublands that remain in scattered locations across the eastern United States. These open areas, with their distinctive flora and fauna of wildflowers, grasses, butterflies, and birds that rely on large open habitats and frequent disturbance, were largely produced through the agricultural activity of cutting, burning, and continued grazing that has generated similar grass, shrub, and heath-covered areas across temperate areas worldwide. In many ways these areas are the product of quite abusive treatment of the native forest vegetation by clear-cutting, fire, and intensive agriculture. As The Nature Conservancy, Audubon Society, National Park Service, and numerous state agencies have learned, maintaining these highly desirable cultural landscapes is very difficult without the "free" labor of agriculture and the unintended consequences of the traditional land-use practices that generated them. Consequently, land managers operating in a landscape in which agricultural habitats are desired but farmers are scarce have resorted to prescribed fires, brushwhacking with mechanical harvesters, mowing, and even the importation of sheep in an effort to retain the open, treeless character of such areas as the heathlands of Nantucket, the shrub and grasslands of Martha's Vineyard, Block Island, and Cape Cod, and meadow areas throughout the Northeast. However, as a look at the shrubby and woody state of most of Thoreau's formerly grass-covered meadows will show, it is seldom possible to keep the trees and shrubs out of any extensive areas without constant maintenance and heroic effort. It is ironic that some of the best examples of openland habitat and the largest "preserves" of grassland in Massachusetts now exist on completely artificial landscapes that received extremely harsh treatment in the past, including airfields, landfills, military bases, and military training reserves. On these sites regular mowing controls the woody vegetation, and fires, explosives, and heavy equipment provide the disturbance necessary to enable open-habitat species to persist. Thus, in an effort to maintain some of our distinctive "native" birds, butterflies, and plants that depend on open habitats, conservationists find themselves resorting to highly artificial means or extolling the virtues of artificial habitats. We are caught in a cultural dilemma in which we seek to maintain what we know and what is becoming rare even though it is largely the consequence of intense human activity.

In our effort to understand natural history, whether for enjoyment or as an applied or academic enterprise, we seldom give enough attention to reviewing what has been learned in the past. This can lead to the expenditure of considerable time rediscovering what is already known. An example is the fruitless efforts of a generation of foresters who tried to encourage white pine (a valuable timber species) rather than hardwoods (generally of lower value) to succeed the cutting of the old-field white pine forests. If these foresters had read and understood Thoreau's "Succession of Forest Trees" and other writings, they would have recognized that the white pine forest was quite "unnatural" as a dominant forest type in most of the New England uplands and that the hardwoods would generally outcompete pine despite the best human efforts. Unfortunately, it took a generation of frustrating attempts at regenerating pine on sites that became dominated by hardwoods, as well as intensive research by many students and faculty, to acknowledge the futility of these efforts and to (re)discover the historical and biological facts about white pine forests in New England. Harvard University's Fisher Museum at the Harvard Forest has a series of exhibits that depict land-use history and forest change in New England. Although the museum basically tells the story of Thoreau's "Succession of Forest Trees," the actual exhibits and dioramas, which were designed by Harvard faculty and crafted in the 1930s, do not mention Henry Thoreau. More than half a century after Thoreau laid out the story of succession in painstaking detail in his journals, his lessons had to be relearned by the forest ecologists at Harvard.

A second example of the value of studying Thoreau's observations and learning from the past involves a project I worked on with Glenn Motzkin at the Harvard Forest for nearly two years. We were seeking to interpret the patterns of vegetation on the largest sand plain remaining in the Connecticut River Valley. This site was of great conservation interest because of the abundance of pitch pine and scrub oak vegetation, which is uncommon and supports a number of unusual plant and animal species. Coincidentally, in the evenings during this period I was reading Thoreau's journals, which contain many observations on these species and vegetation types. At that point in our work it was clear to us that pitch pine occurred predominantly on sites that had been plowed in the distant past and then abandoned from agriculture, as indicated by a one-foot-deep "plow layer" that was quite obvious in the sandy soil. In contrast, the scrub oak vegetation formed an abrupt transition with the pines on adjacent sites that were lacking this distinctive soil horizon. From this observation we deduced that the land containing scrub oak must never have been cleared or farmed. In all other features except this detail of history, including slope, soil texture, and fertility, the environment and sites that supported these very distinctive vegetation types were identical. I then discovered, with a bit of chagrin mixed with interest, that Thoreau had frequently observed a similar pattern in his walks. In his journals he provided a clear interpretation of this striking arrangement of vegetation:

> Observed to-day on the edge of a wood-lot of Loring's, where his shrub [scrub] oaks bounded on a neighbor's small pitch pines, which grew very close together, that the line of separation was remarkably straight and distinct, neither a shrub oak nor a pine passing its limit, the ground where the pines grew having apparently been cultivated so far, and its edges defined by the plow.
>
> OCTOBER 19, 1851

Our initial interpretations were corroborated by this passage: the sharp break between the oak and pine vegetation types was indeed controlled by land-use and ownership differences. In addition, Thoreau's observations convinced us that the phenomenon was potentially quite widespread on sandy soils elsewhere in New England. Thoreau also answered a question that we had raised but could never hope to resolve even through detailed historical reconstructions: what was the explicit use of such scrub oak areas, and how did they arise? According to his observations the pine area had been plowed, as we had confirmed, but the scrub oak area had been a woodlot. On the basis of this information we deduced that the tree oaks had been cut so many times, and perhaps burned, that a shrubby landscape of gnarly scrub oak had developed. Our further investigations showed that this pattern holds broadly for pitch pine stands on coarse sandy soils throughout the Connecticut River Valley.

Thus we were able to relate the modern characteristics of the vegetation and the patterns they form in the landscape to the combination of environmental conditions and past land use. These conclusions shed light on the problems that exist for conservationists who are seeking to retain or restore the uncommon assemblages of plants and animals that inhabit pitch pine and scrub oak vegetation. Because these and many other distinctive vegetation types occur as a consequence of specific historical land-use practices, they will continue to change into other forest types and will gradually disappear from the land. Recognition of this potential for change and loss does not provide simple answers for conservationists, but it greatly clarifies the scope of the problem and possible management options. Our inclusion of Thoreau's words in our paper published 136 years after his initial observations is an acknowledgment of the value of looking back in time.

It is instructive to note that there is a major contradiction between Thoreau's landscape and writing, on the one hand, and his modern image, on the other. Thoreau is often perceived and cited by twentieth-century readers as the proponent of a wilderness ethic that led to a strongly preservationist sentiment. Indeed, his words are printed on countless nature posters, calendars, and T-shirts depicting scenes of wilderness landscapes, old-growth forests, and deep woods. Thoreau's name alone signifies conservation with an emphasis on preservation. Nevertheless, Thoreau actually lived in a landscape where the woods were relatively few and heavily cut, where fields and farms predominated, and where people were actively and incessantly working the entire countryside for all available natural resources. Although he railed occasionally against the tameness of Concord's nature and dreamed of the wilder landscape of the colonists, in fact Thoreau thoroughly embraced his agricultural landscape and its people and spent much of his time in the forests conceptualizing plans for the sound management of woodlots.

Despite the cleared forests, the dwindling animal populations, the dammed and polluted rivers, and the declining numbers of waterfowl and fish, Thoreau was able to find wildness in a thousand scenes, each one shaped by human activity. Understanding full well the cultural origins of young pine forests on abandoned pasturelands, he could still remark that "these are our loveliest woods." And, of course, he could turn Walden, a cut-over and "tamed" woodlot, whose shores had recently been desecrated by one thousand workers building the railroad to Fitchburg, into a symbol of solitude, natural values, and wilderness.

This apparent contradiction leaves us with two ideas to ponder. The first is that wilderness and perhaps all possible experiences in life can be found inside oneself. This was an overriding conviction of Thoreau's and a basis for his life and writing. Although he saw many sites across New England and went as far afield as the Mississippi River on his Minnesota voyage, he seldom felt the urge to leave Concord because he felt that he could find nearly every natural scene and experience right there. In fact, it is remarkable how well his natural and cultural history observations fit the broader landscape of the northeastern United States. The second idea, which is relevant to both our perception and our management of nature, is that we live in a cultural landscape that we can appreciate for its natural qualities as well as for the full story of human history that it contains. Recognition that almost all landscapes are shaped by the interaction of several things—the environment, plants and animals, and people—helps us to interpret their natural history. It may also prevent us from attempting to identify and preserve a wilderness or nature that is untouched by us, outside of us, or a remnant of "original conditions." Thoreau could describe a swamp or a woods in Concord as "wild as the wildest scene in Labrador" while fully recognizing the role that people had played in shaping it into what it was. He also appreciated fully the scenes created by people. His journals thus contain not the writings of a wilderness retreat, but an evocation of the range of moods, emotions, and natural history observations that come from an old and humanized landscape like that of New England. We can benefit from Thoreau's recognition that every landscape has been touched by people, and we can use his approach to appreciate, understand, and conserve our countryside today.

Bibliographic Essay

Living in solitude in the deep woods teaches one a sense of humility before the raw and often uncaring side of nature. This lesson was brought home to Thoreau when he traveled to Maine; in describing his experience on Mount Katahdin, he wrote: "Perhaps I most fully realized that this was the primitive, untamed, and forever untameable *Nature,* or whatever else men call it, while coming down this part of the mountain . . . Nature was here something savage and awful, though beautiful." In fact, as noted by Thoreau's biographers, R. Richardson and W. Harding, he thrived in the cultural landscape of New England, where towns and farms coexisted with swamps, meadows, and woodlots, and where Thoreau could move easily from a fence-side conversation with a local farmer or woodchopper to an animated exchange with Bronson Alcott or Nathaniel Hawthorne in Ralph Waldo Emerson's study, and then on to a relatively peaceful retreat alongside Walden Pond.

It is clear that Thoreau not only recognized the extent to which his landscape was humanized, but he actually attributed his inspiration and artistry to the civilized character of the land. On returning from the Maine woods, he remarked: "Nevertheless, it was a relief to get back to our smooth, but still varied landscape. For a permanent residence, it seemed to me that there could be no comparison between this [Concord] and the wilderness. The wilderness is simple, almost to barrenness. The partially cultivated country it is which chiefly has inspired, and will continue to inspire, the strains of the poets." Thoreau's keen interest in the human character of his landscape is well captured in F. H. Allen's *Men of Concord and Some Others,* which is based on entries from Thoreau's journals and illustrated with watercolors of N. C. Wyeth. This strongly humanized aspect of Thoreau's landscape is evident throughout his journals.

Although Thoreau was one of the most observant and prolific writers to document the land, natural history, and human activity of his time, few of his detailed observations and insightful reflections on natural history were actually included in his published writings, such as *Walden* or *A Week on the Concord and Merrimack Rivers.* These two books were so carefully constructed and extensively edited that they were lacking much of the ecological insight that characterized the journals. It is known that although Thoreau had a first draft of *Walden* completed in September 1847, it took nearly seven years and eight complete revisions before the book emerged in its final form. By that time almost none of the lyrical, whimsical, and detailed observations on nature that are contained in the actual journal entries from his Walden Pond days remained in the text. Even Thoreau's published paper entitled "Natural History in Massachusetts" does not present a compilation of his own observations on nature or even many interesting ecological insights. Rather, this particular piece was prompted by Emerson's suggestion that Thoreau provide a review in an 1842 volume of the literary journal *The Dial* on several scientific surveys that had been published on the flora, fauna, and natural resources of Massachusetts. It is precisely because Thoreau's journals provide remarkable, wide-ranging, and insightful observations on everyday sights and activity in the nineteenth-century landscape that I have drawn all but a few of the quotes in this book from that source.

Composition and Publication of the Journals

During his lifetime Thoreau maintained two major sets of writings: his journals, which eventually filled over 30 volumes with nearly two million words, and a collection of commonplace books that constitute his "Fact Book," in which he copied quotes and ideas that totaled nearly 6,000 pages (see K. W. Cameron's transcription and contextual information on Thoreau's notes in *Thoreau's Fact Book*). As part of this prodigious note-keeping, Thoreau developed a collection of eleven notebooks pertaining to Indian history, lore, and characteristics that have become known as the "Indian Book." Both Thoreau's journals and his fact books served as sources for his lectures, essays, and books. The journals were maintained on nearly a daily basis and represent his lifelong work—a platform for developing his skills of observation and natural history interpretation, and his major literary accomplishment. After Thoreau's death the journals passed to his sister Sophia and ultimately to his correspondent and disciple H. G. O. Blake in Worcester, Massachusetts. Although volumes of excerpts were subsequently produced in various forms, including a seasonal arrangement edited by Blake, the journals were first compiled and published by Bradford Torrey and Francis H. Allen in 1906, accompanied by a number of Herbert W. Gleason's superb photographs of Thoreau's Concord. A reprinted version of Torrey and Allen was released by Dover Publications in 1962 and provided the source material for this book. The journals are currently being completely reedited, including some material lacking in the Torrey and Allen version, by Princeton University Press (J. C. Broderick, general editor), and five volumes through 1853 are now available.

In selecting from Thoreau's writings for this book I first compiled, in chronological order, quotes from the journals that illustrated landscape scenes, natural history processes, or land-use activities that provided me with new or refined insights into the history and ecology of New England, along with a few other entries that were simply amusing or delightful. I then organized this collection into topics corresponding to the present chapters and reduced the material to a small subset for inclusion in the final book. The quotes were transcribed from the Torrey and Allen edition of the journals; interested readers should refer to the Princeton volumes for a discussion of differences between the original journals and various published versions. Two consistent modifications have been made to the entries included in this book: I used ellipses to shorten quotes in ways that focused them without altering their meaning, and I added comments in square brackets (in addition to the few editorial comments or corrections inserted by Torrey and Allen, also in square brackets) to explain specific terms or Latin names employed by Thoreau. All comments in standard parentheses are contained in the original by Thoreau, as reproduced by Torrey and Allen.

In arranging the journal entries at the end of each chapter I have organized them either seasonally or topically in order to form logical groups. Although literary scholars have noted that the style and subject of Thoreau's journal entries changed through his lifetime, basically from a philosophical and lyrical style to a more scientific quality as he grew older, I have deviated from a strict chronological organization of the quotes in order to focus on natural history observations and insights rather than on the personal development of the writer.

The Cultural Landscape of New England

In putting together this book, Abby Rorer and I worked together closely to develop illustrations that would capture the spirit of the different sections of the book and insights into nature rather than portray specific land-use practices or exact observations made by Thoreau. I have chosen not to include maps of specific locations in Concord or places that Thoreau visited elsewhere in New England for two reasons. First, my intention in developing this volume has been to illustrate the broad connection between land-use history and modern landscape characteristics and thereby to emphasize the general applicability of Thoreau's descriptions to a large part of eastern North America, rather than to root the observations within Concord or even New England. Second, as pointed out by Lawrence Buell in *The Environmental Imagination,* Thoreau's descriptions, although detailed and site-specific, actually rely very little on geographic context.

However, for a detailed portrayal of Thoreau's Concord landscape, readers can turn to Herbert Gleason's turn-of-the century photographs contained in *Thoreau Country* (M. Silber, ed.) or to Gleason's own *Through the Year with Thoreau.* The extent and actual route of Thoreau's various travels are well mapped in Stowell's *A Thoreau Gazetteer,* and specific details on his last trip to the midwestern United States are given by W. Harding in *Thoreau's Minnesota Journey.* Published issues of the *Thoreau Society Bulletin* and an astonishingly large number of Web sites can provide the interested Thoreauvian with almost unlimited options for further investigation into the man, his life, his travels, and more. In its new home in Lincoln, Massachusetts, the Thoreau Institute will provide a major resource of archives, artifacts, and literature that will be useful to all researchers of Thoreau.

The remarkable changes that have occurred in the economy, culture, and landscape features of New England have provided a rich source of scholarship for historians, economists, and geographers for more than a century, and there is an extensive literature to explore. The contrasts in culture and landscape between Indian and European times are well portrayed in William Cronon's *Changes in the Land* and Carolyn Merchant's *Ecological Revolutions.* Cronon uses many of the early historical sources that were familiar to Thoreau to argue, as Thoreau did, that Indian culture and history exerted a pronounced effect on the forests and landscape of New England. This argument first achieved widespread recognition in ecological circles through the publication of Gordon Day's article "The Indian as an

Ecological Factor in the Northeast." Merchant brings the prehistory of New England through what she terms a second "ecological revolution," that of capitalist industrialization, which led to a decline in agriculture, concentration of the population into cities, and subsequent reforestation of the countryside. A superb ecological overview of the last 500 years of change in a data-rich companion to these more general, historical treatments is provided by Gordon Whitney's *From Coastal Wilderness to Fruited Plain.* With more than 2000 bibliographic references, considerable original data, synthetic figures, and broad coverage of the eastern United States, Whitney's book provides an unparalleled source of information on the ecological history of the entire northeastern region.

The landscape history of New England and Thoreau's process of "forest succession" are conveyed in a beautiful and accurate series of dioramas and through self-guided nature trails at Harvard University's Fisher Museum located at the Harvard Forest in Petersham, Massachusetts. An interpretation of the dioramas and the importance of this history to ecology, conservation, and forest management is provided in a booklet published by the Harvard Forest (Foster and O'Keefe) and in Hugh Raup's classic paper "The View from John Sanderson's Farm." The latter traces the changes in the land through a focus on the history of three generations of one family that settled in central Massachusetts and developed a colonial farmstead. In my own publications I have sought to document, explore, and discuss the ecological and conservation implications of this history with regard to individual woods, contrasting landscapes across Massachusetts, and the entire New England region. For an interesting companion to the story of forests across the uplands, readers should consult the work of Peter Dunwiddie, former ecologist with the Massachusetts Audubon Society, who has paired historical and modern photographs from the same locations in his interpretation of landscape change on Nantucket, Martha's Vineyard, and Cape Cod.

The broad history of deforestation, agricultural development and abandonment, and reforestation outlined in Thoreau's writings and this volume holds for much of the eastern United States and is reviewed in articles by G. G. Whitney, M. Williams, D. MacCleery, and L. Irland. In New England this pattern of landscape history pertains to the entire region outside of northern Maine and the northernmost and most mountainous parts of New Hampshire and Vermont. In support of this last statement, readers should consult E. L. Bogart's *Peacham,* a detailed overview of one hilltown in the Northeast Kingdom of Vermont; P. W. Gates's "Two hundred years of farming in Gilsum (New Hampshire)"; A. K. Botts's "Northbridge, Massachusetts, a town that moved down hill"; R. J. Favretti's *Highlights of Connecticut Agriculture;* and W. F. Robinson's *Mountain New England: Life Past and Present.* Although dealing with quite dissimilar parts of New England, each of these works portrays a history that is remarkably similar to that captured in the Fisher Museum dioramas.

The notion of a "cultural landscape," which is referred to throughout the text, emerged in Scandinavia in the 1930s; the ecological and historical background of this concept is well described by Knut Faegri from Bergen and other authors in *The Cultural Landscape: Past, Present and Future,* edited by H. Birks and colleagues. For much of the European continent, which has been shaped by human activity at least since the early Neolithic Period (approximately 5000 years ago), the concept of a cultural landscape is quite appropriate and has important applications in the areas of conservation biology and land management. I was first exposed to the concept on a field trip across western Norway led by J. Birks, P.-E. Kaland, M. Kvamme, and D. Moe, which accompanied the symposium for the *Cultural Landscape* volume.

The writings of the British ecologists George Peterken and Oliver Rackham convey the historical research that underlies the interpretation of cultural landscapes and highlight some of the ecological lessons and environmental management issues that emerge through an awareness of the cultural element in nature. For many Americans, who hold wilderness and the preservation of wild or primitive states of nature in high esteem, the notion of the cultural landscape and its preservation may seem foreign or even counterproductive. However, ecologists, geographers, and archaeologists are increasingly recognizing that few spots on earth have escaped human impact and that many elements of nature that we prize are, in fact, the products of past cultural activity (see, for example, W. Denevan's "The Pristine Myth: The Landscape of the Americas in 1492," A. Roosevelt's *Amazonian Indians from Prehistory to the Present,* and A. Goudie's *The Human Impact: Man's Role in Environmental Change*).

Views of the Nineteenth-Century Countryside

The landscape and cultural activities of Thoreau's time are portrayed in work by Winslow Homer; in the late nineteenth and early twentieth century photographs by Gleason in *Thoreau Country;* and in the work of Currier and Ives and other contemporary artists. *The Graphic Art of Winslow Homer* (W. Goodrich) is a particularly interesting reference because Homer's etchings, lithographs, and engravings depict many views of hunting, work in the woods, and daily life in villages and farms that are not portrayed in his well-known oil paintings and watercolors. The influence of Homer's work can be seen in all the illustrations for this book, and the illustration of children gathering chestnuts is based directly on one of Homer's drawings.

One of the most informative perspectives on the condition of the New England landscape shortly before Thoreau's time comes from *Travels in New England and New York* by Timothy Dwight, president of Yale College. Dwight traveled extensively throughout the region and provides a thorough description of the landscape and people as well as the customs and activities of the day.

Daily Life

The day-to-day life of agricultural New England was recorded through the common practice of daily journal and account-book entries that were maintained by thousands of individuals across the rural landscape in the eighteenth and nineteenth centuries. These journals have been mined by scholars as sources of information on cultural practices and technological developments and have even been analyzed through a rigorous semi-quantitative application in order to document long-term changes in climate (see W. R. Baron and D. C. Smith). Although many journals survive in archives, libraries, and museums such as Old Sturbridge Village in Sturbridge, Massachusetts, and the American Antiquarian Society in Worcester, few are published and easily accessible. One informative exception is *Very Poor and of a Lo Make: The Journals of Abner Sanger, 1774–1782* (L. Stabler, ed.). Like most journals of the time, Sanger's does not provide the breadth of natural history observations, landscape perspectives, and introspection found in the writings of Thoreau. However, it does offer detailed information on the range of daily activities and chores encountered in rural life, records on the extensive travels of everyday people, and a sense of the life styles and seasonal changes in the New England countryside.

For an extremely informative picture of daily life in the mid-nineteenth century, Old Sturbridge Village is unparalleled in its authentic restoration and portrayal of town and country settings. It provides a marvelous view of New England in 1840, at the peak of its agrarian splendor. The Plimoth Plantation in eastern Massachusetts offers an equally insightful view of a different life style at the time of English settlement in a coastal village in the 1620s.

In depicting daily life in the mid-nineteenth century I have included a number of quotes touching on Thoreau's home life. Thoreau was a staunch, though very independent, abolitionist, and his emotional involvement in helping slaves flee northward and his support of John Brown emerge vividly in his journals and provide insight into some of the powerful cultural forces that were changing the nation in his day. These issues are explored in more detail in W. Glick's *Henry D. Thoreau: Reform Papers,* as well as in the many historical studies of Thoreau's life, including Harding's and Richardson's.

The Farmer as Hero

In large part because of *Walden,* which portrayed farmers as scarecrows burdened with debt and ignorant even of the better qualities of their own land, Thoreau is often thought to be intolerant of the agricultural life style and his neighbors who pursued it. However, his journals provide another perspective—one that suggests that he not only enjoyed but respected the information and insights from farmers and that he was truly amazed by the extent and results of their activities. His use of terms such as "heroic" and "certain moral worth" for the local farmers, and his comparison of their efforts to those of their Greek and Roman counterparts, pay homage to their accomplishments in developing the agrarian landscape of the eastern United States. Thoreau's attitude toward contemporary agriculture is also reflected in his writing style: he uses classical metaphors and often describes everyday sights and activities in georgic terms (see Lawrence Buell's *The Environmental Imagination*).

Detailed assessments of the regional agricultural economy and history of New England are found in M. Pabst's "Agricultural Trends in the Connecticut River Valley Region of Massachusetts, 1800–1900," W. Rothenberg's *The Market and Massachusetts Farmers, 1750–1855,* and P. Munyon's *A Reassessment of New England's Agriculture in the Last Thirty Years of the Nineteenth Century.* General and more wide-ranging overviews include P. Bidwell and J. Falconer, *History of Agriculture in the Northern United States, 1620–1860;* C. Danhoff, *Change in Agriculture: The Northern United States, 1820–1870;* and J. Black, *The Rural Economy of New England.* A recent interpretation that emphasizes the productivity of the land and the success of farming across the region is provided by M. Bell in "Did New England Go Downhill?" H. Barron emphasizes the fact that many New England farmers chose not to abandon their farms in the nineteenth century, and he offers a delightful view into the subsequent history of the remaining farm families in a small Vermont hilltown in *Those Who Stayed Behind.* For Concord, Massachusetts, the students of David Fischer at Brandeis University present a comprehensive overview and detailed analysis of economic, social, and ecological history in *Concord: The Social History of a New England Town, 1750–1850.*

Meadows and Mowers

The loss of open marshes, streamside and floodplain meadows, and upland fields to forest succession, dam construction, or housing and industrial development constitutes one of the most striking and ecologically important changes in the landscape of most temperate regions of the world from Thoreau's time to ours. The importance of open grass, sedge, and shrub-dominated areas as wildlife habitat and the historical decline of open-land birds, insects, and mammals that has occurred in eastern North America as these areas have decreased are well described in articles by Robert Askins and in *Native Grasslands of the Northeastern United States,* edited by P. Vickery and P. Dunwiddie. In the past decade meadows and fields have emerged as a major focus for many conservation organizations, and these efforts are well described in publications from the Massachusetts Audubon Society.

Historical management of the Concord meadows and the nature of the conflicts that developed between the industrialists who owned the dams and thereby controlled the height of riverflow and the farmers who worked the meadows are the focus of B. Donohue's "Damned at Both Ends and Cursed in the Middle: The Flowage of the Concord River, 1798–1862." Today these former grassy areas in Concord are nearly all covered with shrublands or trees. Although they flood annually in the spring, the old meadows bear slight resemblance to the grassy rivers described in the mid-1800s by Thoreau.

Stone Walls and Other Fences

Those interested in exploring the stone walls, cellar holes, and other relics of New England's agricultural past need only find a country lane or rural wooded area to begin their quest. One superb location for such pursuits is the forest surrounding the Quabbin Reservoir in north central Massachusetts, where more than 70,000 acres of land are protected and managed by the Commonwealth of Massachusetts to shelter the watershed that provides Metropolitan Boston's drinking water. Since the 1930s, when the existing towns were disincorporated and the houses were razed to create the Quabbin Reservoir, the land has experienced little use other than forestry and recreation. Consequently, the layout of roads and village centers and a wide range of stone fences, foundations, and other artifacts have been preserved in their historical setting, albeit with a new forest surrounding them. T. Conuel's *Quabbin: The Accidental Wilderness* provides a wonderful introduction and historical perspective to this interesting landscape.

The dioramas, exhibits, and walking trails at the Fisher Museum of the Harvard Forest in neighboring Petersham provide another good source of information about the patterns and types of stone walls and their relationship to different land-use practices. An abundance of facts and lore about stones and stone walls are contained in Susan Allport's *Sermons in Stone.* Neil Jorgensen's delightful *A Guide to New England's Landscape* and *A Sierra Club Naturalist's Guide to Southern New England* provide general information on characteristics and clues for reading the New England landscape. Michael Bell's article "Did New England Go Downhill?" presents an interesting assessment of fences and stone walls in relation to the agricultural history of the region, but his attempt to downplay the ubiquity and importance of stone walls in New England is not supported by an examination of the landscape itself or maps such as the one compiled for Petersham, Massachusetts, located in the Fisher Museum. Illustrations of other fence types and many other agricultural implements are found in Eric Sloane's books, and a broad overview of artifacts from New England's colonial and agricultural history is presented in W. Robinson's *Abandoned New England.*

Woodlands and Sproutlands

An informative assessment of Thoreau's woodland classification and the relationship of the different forests to the agricultural history of Concord is presented by G. Whitney and W. Davis in "From Primitive Woods to Cultivated Woodlots: Thoreau and the Forest History of Concord, Massachusetts." This subject and the history of the New England landscape are then placed in the broader context of Northern European and North American woodlands and forest practices by G. Peterken in *Natural Woodlands.* Oliver Rackham provides a nice overview of British woodland practices, many of which were imported into North America by early immigrants, in his *Ancient Woodland: Its History, Vegetation and Uses in England.* A work that covers a wide range of natural processes, cultural practices, and historical changes in Massachusetts is *Stepping Back to Look Forward: A History of Massachusetts Forests,* edited by C. H. W. Foster.

George Peterken has mentioned his surprise that Thoreau would interpret pollarding as an Irish custom, since George's considerable experience indicates that pollarding has been quite uncommon in Ireland. Throughout the journals, Thoreau has an interesting habit of attributing many peculiar sights in his landscape to the newly arrived Irish immigrants; it is undoubtedly true in the case of pollards, and perhaps elsewhere, that such assertions and generalizations were occasionally incorrect. Nonetheless, as is clearly shown in Peterken's and Rackham's work and in H. Birks' *Cultural Landscape,* pollarding was common elsewhere in the British Isles and Europe, and it continues to give a distinctive appearance to many parks, hedgerows, and farmyards.

The problem of overgrazing by deer and other browsing animals is a severe one in many forests as wildlife specialists and land managers struggle to achieve a balance between desirable levels for hunting and overpopulation resulting from overzealous protection of deer by the larger public. Given the absence of adequate natural predators, and with control measures generally limited to hunting, fencing, and sterilization, the management of deer populations has become a very contentious issue across much of eastern North America. Discussion of historical changes in deer populations is provided by Aldo Leopold's early studies, and the impact of deer on forest ecosystems is well covered by G. Whitney and D. Marquis in their assessments of the changes in old-growth and second-growth forests of Pennsylvania brought about by deer.

Forest Land Use and Woodland Practices

The extent to which New England's woodlands were cut over and mismanaged in the nineteenth and early twentieth centuries is well described in such works as A. F. Hawes's "New England Forests in Retrospect" and "The Present Condition of Connecticut Forests: A Neglected Resource," H. O. Cook's "Fifty Years a Forester," L. Irland's *Wildlands and Woodlots: The Story of New England's Forests,* A. Chittenden's "Forest Conditions in Northern New Hampshire," J. Toumey's "What Ails New England Forests?" and C. S. Sargent's "Report on the Forests of North America." A poignant assessment of the over-utilization of Massachusetts woodland for fuelwood, which is contemporary with Thoreau's observations, is given in George Emerson's "Report on the Trees and Shrubs Growing Naturally in the Forests of Massachusetts." A broader perspective on timber and fuelwood consumption that covers the history and geography of U.S. forest use is provided by Michael Williams in *Americans and Their Forests: A Historical Geography.*

Unfortunately, Thoreau's recommendation that forest management should be based on an understanding of woodlot history, the ecology of the individual tree species, and a landscape perspective of natural variation is still not followed very successfully in the majority of logging operations across New England. However, the fundamental soundness of Thoreau's approach is clearly underscored in the many silvicultural texts and publications that have been developed in the past century. Among the most informative, from a historical and ecological perspective, are B. Fernow, "Forestry for Farmers"; S. Spurr and A. Cline, "Ecological Forestry in Central New England"; R. Hawley and A. Hawes, *Forestry in New England: A Handbook of Eastern Forest Management;* S. Spurr and B. Barnes, *Forest Ecology;* D. M. Smith, *The Practice of Silviculture;* and M. Beattie et al., *Working with Your Woodlot: A Landowner's Guide.*

Firewood and Other Fuels

Drawing from wide-ranging sources, G. Whitney's *From Coastal Wilderness to Fruited Plain* provides a comprehensive assessment of the impact that America's dependence on firewood had on the extent, structure, and composition of forests. The early colonial period of forest clearance and profligate fuel and timber use is well covered in C. Carroll's *The Timber Economy of Puritan New England,* while M. Williams brings this story up to Thoreau's lifetime in "Clearing the United States Forests: The Pivotal Years, 1818–1860." An interesting perspective on the role of the railroads in generating the perceived wood shortage in the northeastern United States is provided by S. Olsen in "The Depletion Myth: A History of Railroad Use of Timber."

The consequences of fuelwood cutting on the development of second-growth forests in New England are depicted in the Harvard Forest dioramas at the Fisher Museum and are described in the long-term studies of G. Stephens and P. Waggoner, "A Half Century of Natural Transitions in Mixed Hardwood Forests." In a recent study of regional forest change, "Land Use as Broad-Scale Disturbance: Regional Forest Dynamics in Central New England," D. Foster and others document that forest composition has become increasingly homogeneous as a consequence of the history of repeated cutting and forest succession on abandoned farmlands. Notably, species that sprout effectively after cutting and fire, such as red maple, birches, and oaks, have increased greatly across the region, whereas long-lived and shade-tolerant species such as hemlock, beech, and sugar maple have declined.

Wildfire: A Human and Natural Force

A good general introduction to the topic of fire and its ecological effect on forest ecosystems is provided by S. Pyne in *Fire in America: A Cultural History of Wildland and Rural Fire.* The importance of Native American use of fire in controlling the early structure and composition of the landscape of New England was suggested in a classic paper by G. Day, "The Indian as an Ecological Factor in the Northeastern Forest." An interesting contrasting view that provides a critical assessment of the historical sources used by Day, as well as many of those cited by Thoreau a century earlier, is found in E. Russell, "Indian-set Fires in the Forests of the Northeastern United States," and (with R. T. T. Forman), "Indian Burning: The Unlikely Hypothesis."

The role of pre-European fire remains a much debated topic, as is demonstrated by a recent energetic exchange in the *Journal of Ecology* between M. Abrams and J. Clark. Such debates over the status of presettlement landscapes and the factors that control them have much more than academic importance. For example, in the case of fire and the purported role of Indians in shaping the pre-European forests, the interpretation of the abundance of oak trees, open habitat, and frequency of fire feeds directly into current management decisions made by foresters, wildlife agencies, and conservationists across the eastern United States (see, for example, M. Abrams, "Fire and the Development of Oak Forests").

The ecological effects of recent fires on New England forests has been assessed in a number of studies, notably J. Brown, "The Role of Fire in Altering the Species Composition of Forests in Rhode Island"; W. Niering, "The Role of Fire Management in Altering Ecosystems"; and T. Fahey and W. Reiners, "Fire in the Forests of Maine and New Hampshire." Other efforts to interpret the prehistorical frequency and distribution of fire are presented in C. Lorimer, "The Presettlement Forest and Natural Disturbance Cycle of Northeastern Maine"; J. Fuller, "Impact of Human Activity on Regional Forest Composition"; and W. Patterson, "Indian Fires in the Prehistory of New England" (with W. Sassamann) and "Fire and Disease History of Forests" (with A. Backman). The lengthy period of forest recovery that is initiated by fire or other natural disturbance is highlighted in "Long-term Vegetation Dynamics and Disturbance History of a *Tsuga*-Dominated Forest" by D. R. Foster and T. Zebryk.

Social Change and Farm Abandonment in New England

The 150 years that have elapsed since the onset of industrialization, widespread migration, and abandonment of agricultural land in the northeastern United States have allowed the development of many interpretations of the political, economic, and social factors underlying the ecological transformation of the landscape. Among the most informative are J. Black's *The Rural Economy of New England: A Regional Study,* C. Danhof's *Change in Agriculture: The Northern United States, 1820–1870,* and C. Merchant's *Ecological Revolutions: Nature, Gender and Science in New England.* The changes in the population, farming activity, and landscape of Concord viewed by Henry Thoreau were paralleled across the northeastern United States and led to the development of local and state agricultural societies that attempted to promote and improve farming and to many editorials lamenting the changing character of small towns and rural life. The persistent New England town tradition of hosting "Old Home Days" was developed in order to lure the younger generation annually back to their hometown from their new, primarily urban, residences.

The Succession of Forest Trees

Although many Thoreau scholars describe his observations and writings on succession as "his greatest scientific achievement" (W. Harding), there is scant evidence that his work had much impact on the scientists who ultimately produced the classic studies on the subject. (See, however, H. Caswells's "Predator-mediated Coexistence: A Non-equilibrium Model" for a citation of Thoreau's "Succession" as the first work to emphasize the importance of dispersal in plant succession.) Thoreau's term "succession" was, in fact, adopted by ecologists and foresters, although many now favor other words that do not imply an orderly replacement of one group of species by another, as Thoreau clearly intended.

By the early twentieth century the topic of succession had become the focus of one of the longest and most important debates in the history of ecology. Two fairly distinctive views emerged. The "Clementsian" perspective, named after F. Clements, a prominent early ecologist whose reputation has suffered immensely because of his real and perceived views on succession, holds that vegetation changes in relatively predictable patterns following a disturbance such as fire, forest cutting, or field abandonment, as one "community" of organisms is replaced by another. The process of change and community replacement was interpreted by the Clementsian school as driven by changes in the local environment brought about by the growing vegetation. The final stage of this successional process is a "climax" community of long-lived and generally competitive, shade-tolerant species that can be self-replacing over very long periods of time and under stable climatic conditions.

The opposing "individualistic" or "Gleasonian" perspective was promoted by H. A. Gleason, a plant taxonomist and biogeographer, who argued that species respond quite separately and individualistically to the environmental conditions that develop following a disturbance and that the "successional" process is consequently quite varied and often disorderly. As an outcome of individualistic species behavior, vegetation does not form a well-defined or closely adapted "community" but is simply a collection of species that happen to be able to arrive and develop on a particular site at a particular time. Proponents of the Gleasonian perspective spurn the terms "succession" and "climax," and generally argue that the high frequency of environmental change and natural disturbance and the individualistic nature of organisms lead to a highly dynamic and varied pattern of vegetation. Good reviews of the history and intensity of the debate on succession are presented by W. Drury and I. Nisbet in "Succession" and R. McIntosh in *The Background of Ecology: Concept and Theory.*

Although Thoreau described a fairly regular replacement of white pine and pitch pine forest by oak and other hardwood trees through time, which seems to fit the Clementsian paradigm, Thoreau's writings clearly suggest that he would have sided with Gleason in the subsequent debate. The attention that Thoreau gave to the specific behavior and uniqueness of individual species of plants and animals, the emphasis that he placed on the various methods and modes of seed dispersal and plant reproduction, and his fascination with serendipity in the natural world all illustrate his clear awareness of the importance of individual species' response to the environment and the chance conditions that control the establishment and growth of vegetation after disturbance.

The length of time and the extent of futile effort that foresters expended in trying to get white pine to regenerate on sites that it occupied following agriculture abandonment provide ample evidence that many scientists and professionals remained ignorant of Thoreau's clear message about succession. Nowhere in the major published writings on the management recommendations for white pine or second-growth hardwood forests are

Thoreau's observations or terminology invoked extensively. These include such works as R. Fisher, "Second-growth White Pine as Related to the Former Uses of the Land"; H. Graves and G. Pinchot, "The White Pine: A Study with Tables of Volumes and Yield"; and S. Spring, "The Natural Replacement of White Pine on Old Fields in New England."

Animals: From Bobolinks to Bears

Surprisingly, despite the extensive changes in wildlife populations that have occurred during the relatively brief history of the eastern United States, very few studies have tried to document in detail the specific nature of these changes for a number of contrasting species or to elucidate the factors that have driven them. This is undoubtedly a consequence of at least two factors: the difficult problem of obtaining reliable abundance estimates for current wildlife populations, let alone reconstructing their numbers during previous centuries, and the wide range of research skills that are required in order to piece such information together. A number of books outline some of the general dynamics in wildlife populations and in human attitudes toward animals since European colonization, including P. Matthiessen's *Wildlife in America,* W. Hornaday's *Our Vanishing Wildlife,* and D. Allen's *Our Wildlife Legacy.* The northeastern United States is currently undergoing a major change in wildlife abundance and composition, and many of the forest-dwelling species that are either returning and increasing or arriving for the first time have been discussed in various dissertations, books, and articles; see, for example, the recent articles by James Cardoza in the *Massachusetts Sportsman* and J. B. Trefethen's "The Massachusetts Land and Its Wildlife: A History of the Residential and Migratory Game Birds and Mammals." A general review of wildlife in the context of major habitat types can be found in R. M. DeGraff, "New England Wildlife: Habitat, Natural History, and Distribution" and "New England Wildlife: Management of Forested Habitats." DeGraff has also edited a volume with R. Miller, *Conservation of Faunal Diversity in Forested Habitats,* which focuses more on landscape change.

The decline of grassland and shrubland bird species is discussed by Vickery, Dunwiddie, and others in *Native Grassland of Northeastern North America* and is reviewed in the context of the current Massachusetts landscape by C. Leahy, J. H. Mitchell, and T. Conuel in *The Nature of Massachusetts.* In contrast, the emphasis is on promotion and reintroduction of large mammals and interior forest species in recent articles by J. Sayen and others in *The Northern Forest Forum* and *Restore: The North Woods.*

Thoreau's abilities as a scientist and his contributions to our understanding of wildlife biology, limnology, ecology, and natural sciences have been evaluated in a number of works, including F. Allen, *Thoreau on Birds;* E. Deevey, "A Re-examination of Thoreau's 'Walden'"; J. Hildebidle, *Thoreau: A Naturalist's Liberty;* and R. Angelo, "Thoreau as Botanist: An Appreciation and Critique" and *Botanical Index to the Journal of Henry David Thoreau.*

The Passenger Pigeon

Excellent discussions of the distribution, feeding and breeding behavior, and habitat preferences of passenger pigeons are provided in early ornithological treatises such as A. Bent's *Life Histories of North American Gallinaceous Birds,* F. Chapman's *Handbook of Birds of Eastern North America,* and E. H. Forbush's *Birds of Massachusetts and Other New England States.* Other sources dealing exclusively with the passenger pigeon include D. Snyder, *The Passenger Pigeon in New England,* and A. Schorger, *The Passenger Pigeon: Its Natural History and Extinction.* These books also provide good overviews of the factors, including the loss of forest habitat and rapacious hunting activity, that led to the decline and ultimate extinction of the species. Sara Webb explores the role of pigeons in plant dispersal following the last Ice Age in "Potential Role of Passenger Pigeons and Other Vertebrates in the Rapid Holocene Migrations of Nut Trees." Despite the great importance of this species in the forested landscape of eastern North America, there is little ecological literature that considers the natural role of the passenger pigeon in the temperate deciduous forest, or speculates on the consequences of its disappearance from this ecosystem. However, see C. Merchant's *Ecological Revolutions* for such interesting perspectives as the importance of pigeons as a source of phosphorous fertilizer to forests and beavers as a critical factor in meadow formation and landscape diversification.

The American Chestnut

The American Chestnut Foundation in St. Paul, Minnesota, was founded in the 1980s to disseminate information about the species and to promote efforts to breed and develop a blight-resistant strain of chestnut. The foundation also serves as a clearing house for a wide range of information concerning the status of the species in the United States. The ecology and role of the chestnut in the early forests of the northeastern United States are discussed in papers such as those by A. E. Moss, "Chestnut and Its Demise in Connecticut," J. Murdoch, "Chestnut: Its Market in Massachusetts," and E. Russell, "Pre-blight distribution of *Castanea dentata* (Marsh.) Borkh." The very long term dynamics of the chestnut in a New England forest are examined in "Long-Term Vegetation Dynamics and Disturbance History of a *Tsuga*-Dominated Forest in New England" by Foster and Zebryk. This paper documents that the chestnut became abundant in this particular forest following disturbances, such as fire during the pre-European period and clear-cutting of the forest by early settlers.

An interesting perspective on the current distribution and ecology of the chestnut is provided by F. Paillet, "Growth-Form and Ecology of American Chestnut Sprout Clones in Northeastern Massachusetts." The effect of the loss of chestnut on forest composition is assessed by C. Korstian and P. Stickel in "The Natural Replacement of Blight-Killed Chestnut in the Hardwood Forests of the Northeast." Since the onset of the chestnut blight, the Connecticut Agricultural Research Station in Hamden, Connecticut, has been one of the major centers of research on the tree and the fungal disease. This work has continued to the present under the direction of Dr. Sandra Anagnostokis, who is an authority on chestnut biology, the pathology of its fungal disease, and the potential for biological control of the blight.

Reading Forest and Landscape History

There are many historical, botanical, and archaeological approaches that are useful in interpreting the history of a landscape and its vegetation. Most of the major scholarly tools and references are reviewed by G. Whitney, *From Coastal Wilderness to Fruited Plain,* and E. Russell, *People and the Land through Time.* Many clues come from the land itself, and these are best learned by examining almost any forest and proceeding through a process of continual investigation and questioning. An aid in this process is provided by the self-guided trails at the Fisher Museum in Petersham, Massachusetts, and by such popular texts as N. Jorgensen's *A Guide to the New England Landscape* and *A Sierra Club Naturalist's Guide to Southern New England,* Betty Thomson's *The Changing Face of New England,* and Tom Wessels's *Reading the Forested Landscape: A Natural History of New England.*

Landscape Change

Historians and ecologists alike have been fascinated by questions concerning the nature of the "presettlement" forests of the northeastern United

States, and there are many studies that offer interpretations. Among the best are S. Bromley's "The Original Forest Types of Southern New England" and G. G. Whitney's *From Coastal Wilderness to Fruited Plain.* Researchers at the Harvard Forest have provided long-term reconstructions on the scale of an individual forest to the landscape of Massachusetts in a series of articles by Fisher, Spurr, Raup, Gould, Foster, and Motzkin.

Insights into the Ecology and Conservation of the Land

Because trees grow slowly, climate changes gradually, and forest dynamics require many decades or centuries to develop, ecologists have increasingly recognized that the interpretation, management, and conservation of forest ecosystems must be based on long-term understanding of their past dynamics. This approach is reflected in such papers as D. Sprugel, "Disturbance, Equilibrium, and Environmental Variability: What Is 'Natural' Vegetation in a Changing Environment?"; M. Hunter, G. L. Jacobson, and T. Webb, "Paleoecology and the Coarse-Filter Approach to Maintaining Biological Diversity"; and J. Lawton, "The Science and Non-Science of Conservation Biology." The implications of retrospective studies for ecology have been described by D. R. Foster and others in "Insights from Paleoecology to Community Ecology" and "Ecological and Conservation Insights from Retrospective Studies of Old-Growth Forests," and the application of this information to the conservation of New England landscapes has been developed in specific studies by A. Golodetz, "Land Protection in Central New England: Historical Development and Ecological Consequences," G. Motzkin et al., "Controlling Site to Evaluate History: Vegetation Patterns of a New England Sand Plain," D. R. Foster and G. Motzkin, "Ecology and Conservation in the Cultural Landscape of New England: Lessons from Nature's History," and P. W. Dunwiddie in *Changing Landscapes: A Pictorial Field Guide to a Century of Change on Nantucket.*

BIBLIOGRAPHY

Abrams, M. "Fire and the Development of Oak Forests." *BioScience* 42 (1992): 346–353.

Allen, D. L. *Our Wildlife Legacy.* New York: Funk & Wagnalls, 1974.

Allen, F. H. (ed.). *Men of Concord and Some Others as Portrayed in the Journal of Henry David Thoreau.* Boston: Houghton Mifflin, 1936.

——— (ed.). *Thoreau on Birds: Notes on New England Birds from the Journals of Henry David Thoreau.* Boston: Beacon Press, 1993.

Allport, S. *Sermons in Stone: The Stone Walls of New England and New York.* New York: Norton, 1990.

Anagnostakis, S. L. "The American Chestnut: New Hope for a Fallen Giant." Bulletin 777. The Connecticut Agricultural Experiment Station, New Haven, 1978.

——— "Biological Control of Chestnut Blight." *Science* 215 (1982): 466–471.

——— "Chestnut Blight: The Classical Problem of an Introduced Pathogen." *Mycologia* 79 (1987): 23–37.

——— "Chestnuts in Our Forest." *Connecticut Woodlands* (on-line edition), Summer 1997.

Angelo, R. *Botanical Index to the Journal of Henry David Thoreau.* Salt Lake City: Peregrine Smith Books, 1984.

——— "Thoreau as Botanist: An Appreciation and a Critique." *Arnoldia* 45 (1985): 13–23.

Askins, R. A., J. F. Lynch, and R. Greenberg. "Population Declines in Migratory Birds in Eastern North America." *Current Ornithology* 7 (1990): 1–57.

Askins, R. A., and M. J. Philbrick. "Effect of Changes in Regional Forest Abundance on the Decline and Recovery of a Forest Bird Community." *Wilson Bulletin* 99 (1987): 7–21.

Askins, R. A., M. J. Philbrick, and D. S. Sugeno. "Relationship between the Regional Abundance of Forest and the Composition of Forest Bird Communities." *Biological Conservation* 39(1987): 129–152.

Audubon, J. J. *Ornithological Biography* [1840–1844]. New York: Volair Books, 1979.

Baldwin, H. I. "The Induced Timberline of Mount Monadnock, New Hampshire." *Bulletin of the Torrey Botanical Club* 104 (107): 324–333.

Baron, W. R., and D. C. Smith. "Growing Season Parameter Reconstruction for New England Using Killing Frost Records, 1697–1947." Bulletin 4846. Maine Agricultural and Forest Experiment Station, University of Maine, 1996.

Barron, H. S. *Those Who Stayed Behind: Rural Society in Nineteenth-Century New England.* Cambridge: Cambridge University Press, 1984.

Beattie, M., C. Thompson, and L. Levine. *Working with Your Woodland: A Landowner's Guide.* Hanover, N.H.: University Press of New England, 1983.

Bell, M. M. "Did New England Go Downhill?" *Geographical Review* 79 (1989): 450–466.

Bent, A. C. *Life Histories of North American Gallinaceous Birds.* New York: Dover, 1963.

Bidwell, P. W., and J. J. Falconer. "History of Agriculture in the Northern United States, 1620–1860." Carnegie Institute Publication no. 358. New York: Peter Smith, 1941.

Birks, H. H., H. J. B. Birks, P. E. Kaland, and D. Moe (eds.). *The Cultural Landscape: Past, Present, and Future.* Cambridge: Cambridge University Press, 1988.

Black, J. D. *The Rural Economy of New England: A Regional Study.* Cambridge, Mass.: Harvard University Press, 1950.

Blake, H. G. O. *Early Spring in Massachusetts: From the Journal of Henry D. Thoreau.* Boston: Houghton Mifflin, 1881.

——— *Autumn: From the Journal of Henry D. Thoreau.* Boston: Houghton Mifflin, 1892.

Bogart, E. L. *Peacham: The Story of a Vermont Hill Town.* Montpelier: Vermont Historical Society, 1942.

Botts, A. K. "Northbridge, Massachusetts, a Town That Moved Down Hill." *Journal of Geography* 33 (1934): 249–260.

Broderick, J. C. (ed.). *Henry D. Thoreau Journal.* Princeton, N.J.: Princeton University Press, 1984.

Brown, J. H. "The Role of Fire in Altering the Species Composition of Forests in Rhode Island." *Ecology* 41 (1960): 310–316.

Buell, L. *The Environmental Imagination.* Cambridge, Mass.: Harvard University Press, 1995.

Cameron, K. W. *Thoreau's Fact Book. In the Harry Elkins Widener Collection in the Harvard College Library.* Hartford: Transcendental Books, 1966.

Carroll, C. F. *The Timber Economy of Puritan New England.* Providence, R.I.: Brown University Press, 1974.

Caswell, H. "Predator-mediated Coexistence: A Non-equilibrium Model." *American Naturalist* 112 (1974): 127–154.

Chapman, F. M. *Handbook of Birds of Eastern North America.* New York: Appleton, 1922.

Chittenden, A. K. "Forest Conditions of Northern New Hampshire. In *Biennial Report of the [New Hampshire] Forestry Commission for the Years 1903–1904,* 1–131. Concord, N.H.: Rumford Printing Company, 1904.

Clements, E. E. "Plant Succession: An Analysis of the Development of Vegetation." *Carnegie Institution of Washington Publication* 242 (1916): 1–512.

Conuel, T. *Quabbin: The Accidental Wilderness,* rev. ed. Amherst: University of Massachusetts Press, 1990.

Cook, H. O. *Fifty Years a Forester.* New Bedford, Mass.: Massachusetts Forests and Park Association, 1961.

Cronon, W. *Changes in the Land: Indians, Colonists, and the Ecology of New England.* New York: Hill and Wang, 1983.

Danhof, C. H. *Change in Agriculture: The Northern United States, 1820–1870.* Cambridge, Mass.: Harvard University Press, 1969.

Day, G. M. "The Indian as an Ecological Factor in the Northeastern Forest." *Ecology* 34 (1953): 329–346.

Deevey, E. S. "A Re-examination of Thoreau's 'Walden.'" *Quarterly Review of Biology* 17 (1942): 1–11.

DeGraaf, R., and R. Miller. *Conservation of Faunal Diversity in Forested Habitats.* London: Chapman Hall, 1996.

DeGraaf, R. M., and D. D. Rudis. "New England Wildlife: Habitat, Natural History and Distribution." USDA Forest Service General Technical Report NE-108, 1986.

DeGraaf, R. M., M. Yamasakai, W. B. Leak, and J. W. Lanier. "New England Wildlife: Management of Forested Habitats." USDA Forest Service General Technical Report NE-144, 1992.

Del Tredici, P. "The Buried Seeds of *Comptonia Peregrina,* the Sweet Fern." *Bulletin of the Torrey Botanical Club* 104 (1977): 270–275.

Denevan, W. M. "The Pristine Myth: The Landscape of the Americas in 1492." *Annals of the Association of American Geographers* 82 (1992): 369–385.

Donahue, B. "The Forests and Fields of Concord: An Ecological History, 1750–1850." In D. H. Fischer (ed.), *Concord: The Social History of a New England Town, 1750–1850,* pp. 14–63. Waltham, Mass.: Brandeis University, 1983.

——— "Damned at Both Ends and Cursed in the Middle: The Flowage of the Concord River Meadows, 1798–1862." *Environmental Review* 13 (1989): 47–68.

Drury, W. H., and I. C. T. Nisbet. "Succession." *Journal of the Arnold Arboretum* 54 (1973): 331–368.

Dunwiddie, P. W. "Forest and Heath: The Shaping of the Vegetation on Nantucket Island." *Journal of Forest History* 33 (1989): 126–133.

——— *Changing Landscapes: A Pictorial Field Guide to a Century of Change on Nantucket.* New Bedford, Mass.: Nantucket Conservation Foundation, 1992.

Dwight, T. *Travels in New England and New York* [New Haven, 1821–1822]. Cambridge, Mass.: Belknap Press, 1969.

Emerson, G. E. *A Report on the Trees and Shrubs in Massachusetts.* Boston: Dutton and Wentworth, 1846.

Fahey, T. J., and W. A. Reiners. "Fire in the Forests of Maine and New Hampshire." *Bulletin of the Torrey Botanical Club* 108 (1981): 362–373.

Favretti, R. J. *Highlights of Connecticut Agriculture.* Storrs, Conn.: College of Agriculture and Natural Resources, University of Connecticut, 1977.

Fernow, B. E. "Forestry for Farmers." *U.S. Department of Agriculture Yearbook for 1894*, 461–500, 1895.

Fischer, D. H. (ed.). *Concord: The Social History of a New England Town, 1750–1850.* Waltham, Mass.: Brandeis University, 1983.

Fisher, R. T. "Second-growth White Pine as Related to the Former Uses of the Land." *Journal of Forestry* 16 (1918): 253–254.

——— "Our Wildlife and the Changing Forest." *The Sportsman,* March 1929, 35–46.

——— "New England's Forests: Biological Factors." *American Geographical Society,* Special Publication 16 (1933), 213–223.

Forman, R. T. T. *Concord's Mill Brook: Flowing Through Time.* Concord, Mass.: Natural Resources Commission, 1977.

Foster, C. H. W. (ed.). *Stepping Back to Look Forward: A History of Massachusetts Forests.* Petersham, Mass.: Harvard Forest, 1998.

Foster, D. R. "Disturbance History, Community Organization, and Vegetation Dynamics of the Old-Growth Pisgah Forest, South-western New Hampshire, U.S.A." *Journal of Ecology* 76 (1988): 105–134.

——— "Land-Use History (1730–1990) and Vegetation Dynamics in Central New England, U.S.A." *Journal of Ecology* 80 (1992):753–772.

——— "Land-Use History and Transformations of the Forest Landscape of Central New England." In S. T. A. Pickett and M. McDonnell (eds.), *Humans as Components of Ecosystems: Subtle Human Effects and the Ecology of Populated Areas,* pp. 91–110. New York: Springer-Verlag, 1993.

——— "Land-Use History and Four Hundred Years of Vegetation Change in New England." In B. L. Turner (ed.), *Principles, Patterns, and Processes of Land Use Change: Some Legacies of the Columbian Encounter,* pp. 253–319. SCOPE Publication. New York: John Wiley and Sons, 1995.

Foster, D. R., J. Aber, R. Bowden, J. Melillo, and F. Bazzaz. "Forest Response to Disturbance and Anthropogenic Stress." *BioScience* 47 (1997): 437–445.

Foster, D. R., and E. Boose. "Patterns of Forest Damage Resulting from Catastrophic Wind in Central New England, U.S.A." *Journal of Ecology* 80 (1992): 79–98.

——— "Hurricane Disturbance Regimes in Temperate and Tropical Forest Ecosystems." In M. Coutts (ed.), *Wind Effects on Trees, Forests, and Landscapes,* pp. 305–339. Cambridge: Cambridge University Press, 1994.

Foster, D. R., and G. Motzkin. "Ecology and Conservation in the Cultural Landscape of New England: Lessons from Nature's History." *Northeastern Naturalist* 5 (1998): 111–126.

Foster, D. R., G. Motzkin, and B. Slater. "Land-Use History as Long-Term Broad-Scale Disturbance: Regional Forest Dynamics in Central New England." *Ecosystems* 1 (1998): 96–119.

Foster, D. R., and J. O'Keefe. *Scenes from the Harvard Forest Dioramas: Insights into the History, Ecology, and Management of New England Forests.* Harvard Forest, Petersham, Mass. In press.

Foster, D. R., D. A. Orwig, and J. McLachlan. "Ecological and Conservation Insights from Retrospective Studies of Old-Growth Forests." *Trends in Ecology and Evolution* 11 (1996): 419–424.

Foster, D. R., P. K. Schoonmaker, and S. T. A. Pickett. "Insights from Paleoecology to Community Ecology." *Trends in Ecology and Evolution* 5 (1990): 119–122.

Foster, D. R., and T. M. Zebryk. "Long-Term Vegetation Dynamics and Disturbance History of a *Tsuga*-Dominated Forest in New England." *Ecology* 74 (1990): 982–998.

Foster, D. R., T. Zebryk, P. K. Schoonmaker, and A. Lezberg. "Post-Settlement History of Human Land-Use and Vegetation Dynamics of a *Tsuga canadensis* (Hemlock) Woodlot in Central New England." *Journal of Ecology* 80 (1992): 773–786.

Frothingham, E. H. "Second-Growth Hardwoods in Connecticut." U.S.D.A. Forest Service Bulletin 96 (1912). Washington, D.C.: U.S. Government Printing Office, 1912.

Fuller, J. L., D. R. Foster, J. S. McLachlan, and N. Drake. "Impact of Human Activity on Regional Forest Composition and Dynamics in Central New England." *Ecosystems* 1 (1998): 76–95.

Garrison, J. R. *Landscape and Material Life in Franklin County, Massachusetts, 1770–1860.* Knoxville: University of Tennessee Press, 1991.

Gates, P. W. "Two Hundred Years of Farming in Gilsum." *Historical New Hampshire* 23 (1978): 1–24.

Gilpin, W. *Observations Relative Chiefly to Picturesque Beauty, Made in the Year 1772 on Several Parts of England.* London: Blamire Strand, 1788.

Gleason, H. A. "The Individualistic Concept of the Plant Association." *Torrey Botanical Club Bulletin* 53 (1926): 7–26.

Gleason, H. W. *Through the Year with Henry D. Thoreau: Sketches of Nature from the Writings of Henry D. Thoreau, with Corresponding Photographic Illustrations.* Boston: Houghton Mifflin, 1917.

Glick, W. *Henry D. Thoreau: Reform Papers.* Princeton, N.J.: Princeton University Press, 1973.

Golodetz, A., and D. Foster. "Land Protection in Central New England: Historical Development and Ecological Consequences." *Conservation Biology* 11 (1996): 227–235.

Goodrich, L. *The Graphic Art of Winslow Homer.* Washington, D.C.: Smithsonian Institution Press, 1968.

Goudie, A. *The Human Impact: Man's Role in Environmental Change.* Cambridge, Mass.: MIT Press, 1982.

Graves, H. S., and R. T. Fisher. "The Woodlot: A Handbook for Owners of Woodlands in Southern New England." U.S.D.A. Bureau of Forestry, Bulletin No. 42 (1903).

Harding, W. *Thoreau's Minnesota Journey: Two Documents. Thoreau's Notes on the Journey West and the Letters of Horace Mann, Jr.* Geneseo, N.Y.: Thoreau Society Booklet No. 16, 1962.

——— *The Days of Henry Thoreau: A Biography.* Princeton, N.J.: Princeton University Press, 1992.

Hawes, A. F. "New England Forests in Retrospect." *Journal of Forestry* 21 (1923): 209–224.

——— *The Present Condition of Connecticut Forests: A Neglected Resource.* New Britain, Conn.: Record Press, 1933.

Hawley, R. C., and A. F. Hawes. *Forestry in New England: A Handbook of Eastern Forest Management,* 1st ed. New York: John Wiley and Sons, 1912.

Hemond, H. F. "Biogeochemistry of Thoreau's Bog, Concord, Massachusetts." *Ecological Monographs* 50 (1980): 507–526.

Hildebidle, J. *Thoreau: A Naturalist's Liberty.* Cambridge, Mass.: Harvard University Press, 1983.

Hornaday, W. T. "The Destruction of our Birds and Mammals: A Report on the Results of an Inquiry." *Annual Report of the New York Zoological Society* 2 (1898): 77–126.

——— *Our Vanishing Wild Life: Its Extermination and Preservation.* New York: New York Zoological Society, 1913.

Howe, R. H. "Thoreau, the Lichenist." *Guide to Nature* 5 (1945): 17–20.

Hunter, M., G. L. Jacobson, and T. Webb. "Paleoecology and the Coarse-Filter Approach to Maintaining Biological Diversity." *Conservation Biology* 2 (1988): 375–385.

Irland, L. *Wildlands and Woodlots: The Story of New England's Forests.* Hanover, N.H.: University Press of New England, 1982.

Jorgensen, N. *A Guide to New England's Landscape.* Chester, Conn.: Pequot Press, 1977.

——— *A Sierra Club Naturalist's Guide to Southern New England.* San Francisco: Sierra Club Books, 1978.

Korstian, C. F., and P. W. Stickel. "The Natural Replacement of Blight-killed Chestnut in the Hardwood Forests of the Northeast." *Journal of Agricultural Research* 34 (1927): 631–648.

Kulfinski, F. B. "The Effect of Grazing upon Succession as Related to Its Use as a Silvicultural Tool in the Maintenance of White Pine." Master's thesis, University of Massachusetts, Amherst, 1953.

Lawton, J. "The Science and Non-Science of Conservation Biology." *Oikos* 79 (1997): 3–5.

Leahy, C., J. H. Mitchell, and T. Conuel. *The Nature of Massachusetts.* Reading, Mass.: Addison-Wesley, 1996.

Leopold, A. "Deer Irruptions." *Wisconsin Conservation Bulletin* 8 (1943): 3–11.

Leopold, A., L. K. Sowis, and D. L. Spencer. "A Survey of Overpopulated Deer Ranges in the United States." *Journal of Wildlife Management* 11 (1947): 162–177.

Little, S., and E. B. Moore. "The Ecological Role of Prescribed Burns in the Pine-Oak Forests of Southern New Jersey." *Ecology* 30 (1949): 223–233.

Lorimer, C. G. "The Presettlement Forest and Natural Disturbance Cycle of Northeastern Maine." *Ecology* 58 (1977): 139–148.

MacCleery, D. *American Forests: A History of Resiliency and Recovery.* Washington, D.C.: U.S. Department of Agriculture Forest Service, 1992.

Marquis, D. A. "Effect of Deer Browsing on Timber Production in Allegheny Hardwood Forests of Northwestern Pennsylvania." USDA Forest Service Research Paper, no. NE-308 (1981). Broomall, Pa., Northeastern Forest Experiment Station.

Matthiessen, P. *Wildlife in America.* New York: Penguin, 1995.

McIntosh, R. P. *The Background of Ecology: Concept and Theory.* Cambridge: Cambridge University Press, 1985.

McKibben, B. "An Explosion of Green." *The Atlantic Monthly,* April 1995, 61–83.

Meeks, H. A. *Time and Change in Vermont: A Human Geography.* Chester, Conn.: Globe Pequot Press, 1986.

Merchant, C. *Ecological Revolutions: Nature, Gender, and Science in New England.* Chapel Hill, N.C.: University of North Carolina Press, 1989.

Moss, A. E. "Chestnut and Its Demise in Connecticut." *Connecticut Woodlands* 38 (1973): 7–13.

Motzkin, G., D. Foster, A. Allen, and J. Harrod. "Controlling Site to Evaluate History: Vegetation Patterns of a New England Sand Plain." *Ecological Monographs* 66 (1996): 345–365.

Munyon, P. G. *A Reassessment of New England Agriculture in the Last Thirty Years of the Nineteenth Century: New Hampshire, a Case Study.* New York: Arno Press, 1978.

Murdoch, J. *Chestnut: Its Market in Massachusetts.* Boston: Office of the State Forester, Wright & Potter Printers, 1912.

Niering, W. A. "The Role of Fire Management in Altering Ecosystems." In H. A. Mooney, T. M. Bonnicksen, N. L. Christensen, J. E. Lotan, and W. A. Reiners (eds.), *Proceedings of the Conference: Fire Regimes and Ecosystem Properties.* USDA Forest Service Technical Report WO-26 (1981): 489–510. Washington, D.C.: USDA Forest Service.

O'Keefe, J., and D. R. Foster. "Ecological History of Massachusetts Forests." In C. H. W. Foster, *Stepping Back to Look Forward: A History of the Forests of Massachusetts,* pp. 19–66. Petersham, Mass.: Harvard Forest, 1998.

Olson, S. H. *The Depletion Myth: A History of Railroad Use of Timber.* Cambridge, Mass.: Harvard University Press, 1971.

Orwig, D. A., and D. R. Foster. "Forest Response to Introduced Hemlock Woolly Adelgid in Southern New England, U.S.A." *Bulletin of the Torrey Botanical Club* 125 (1998): 59–72.

Pabst, M. R. "Agricultural Trends in the Connecticut Valley Region of Massachusetts, 1800–1900." *Smith College Studies in History* 26 (1941): 1–135.

Paillet, F. L. "The Ecological Significance of American Chestnut (*Castanea dentata* [Marsh.] Borkh.) in the Holocene Forests of Connecticut." *Bulletin of the Torrey Botanical Club* 109 (1982): 457–473.

Patterson, W. A., and A. E. Backman. "Fire and Disease History of Forests." In B. Huntley and T. Webb III (eds.), *Vegetation History,* pp. 603–632. Dordrecht: Kluwer, 1988.

Patterson, W. A., and K. E. Sassaman. "Indian Fires in the Prehistory of New England." In G. P. Nichols (ed.), *Holocene Human Ecology in Northeastern North America,* pp. 107–135. New York: Plenum, 1988.

Peterken, G. F. *Woodland Conservation and Management,* 2nd ed. New York: Chapman & Hall, 1993.

——— *Natural Woodland: Ecology and Conservation in Northern Temperate Regions.* Cambridge: Cambridge University Press, 1996.

Pinchot, G. "A Primer of Forestry." U.S. Department of Agriculture, Division of Forestry Bulletin no. 24 (1899).

Pinchot, G., and H. S. Graves. *The White Pine.* New York: Century, 1896.

Pyne, S. *Fire in America: A Cultural History of Wildland and Rural Fire.* Princeton, N.J.: Princeton University Press, 1982.

Rackham, O. "Historical Studies and Woodland Conservation." *Symposium of the British Ecological Society* 11 (1971): 563–580.

——— *Ancient Woodland: Its History, Vegetation and Uses in England.* London: Edward Arnold, 1980.

——— *The History of the Countryside.* London: J. M. Dent, 1986.

Raup, H. M. "Old Field Forests of Southeastern New England." *Journal of the Arnold Arboretum* 21 (1940): 266–273.

——— "The View from John Sanderson's Farm." *Forest History* 10 (1966): 2–11.

Richardson, R. D. *Henry Thoreau: A Life of the Mind.* Berkeley: University of California Press, 1986.

Robinson, W. F. *Abandoned New England: Its Hidden Ruins and Where to Find Them.* Boston: New York Graphic Society, 1976.

——— *Mountain New England: Life Past and Present.* Boston: Little, Brown, 1988.

Roosevelt, A. C. *Amazonian Indians from Prehistory to the Present: Anthropological Perspectives.* Tucson: University of Arizona Press, 1994.

Rothenberg, W. B. "The Market and Massachusetts Farmers, 1750–1855." *Journal of Economic History* 41 (1981): 283–314.

Russell, E. W. B. "Indian-set Fires in the Forests of the Northeastern United States." *Ecology* 64 (1983): 78–88.

——— "Pre-Blight Distribution of *Castanea dentata* (Marsh.) Borkh." *Bulletin of the Torrey Botanical Club* 114 (1987): 183–190.

——— *People and the Land Through Time: Linking Ecology and History.* New Haven: Yale University Press, 1997.

Russell, E. W. B., and R. T. T. Forman. "Indian Burning: The Unlikely Hypothesis." *Bulletin of the Ecological Society of America* 65 (1984): 281–282.

Sargent, C. S. *Report on the Forests of North America.* Washington, D.C.: 10th Census, U.S. Department of Interior, 1884.

Shepard, O. (ed.). *The Heart of Thoreau's Journals.* Boston: Houghton Mifflin, 1927.

Silber, M. (ed.). *Thoreau Country,* with photographs by H. W. Gleason. San Francisco: Sierra Club Books, 1975.

Sloane, E. *Our Vanishing Landscape.* New York: W. Funk, 1955.

——— *American Yesterday.* New York: W. Funk, 1956.

Smith, D. M. *The Practice of Silviculture,* 8th ed. New York: John Wiley and Sons, 1986.

Spring, S. N. "The Natural Replacement of White Pine on Old Fields in New England." U.S.D.A. Bureau of Forestry Bulletin No. 63 (1905).

Sprugel, D. "Disturbance, Equilibrium, and Environmental Variability: What Is 'Natural' Vegetation in a Changing Environment?" *Biological Conservation* 58 (1991): 1–18.

Spurr, S. H., and B. V. Barnes. *Forest Ecology,* 3rd ed. New York: John Wiley and Sons, 1980.

Spurr, S. H., and A. C. Cline. "Ecological Forestry in Central New England." *Journal of Forestry* 40 (1942): 418–420.

Stabler, L. K. *Very Poor and of a Lo Make: The Journal of Abner Sanger, 1774–1782.* Portsmouth, N.H.: Randall, 1986.

Stephens, G. R., and P. Waggoner. "A Half Century of Natural Transitions in Mixed Hardwood Forests." Connecticut Agricultural Experiment Station Bulletin No. 783 (1980).

Stowell, R. F. *A Thoreau Gazetteer.* Princeton, N.J.: Princeton University Press, 1970.

Thomson, B. F. *The Changing Face of New England.* New York: Macmillan, 1958.

Thoreau, H. D. *Faith in a Seed: The Dispersion of Seeds and Other Late Natural History Writings,* ed. B. P. Dean. Washington, D.C: Island Press, 1993.

——— *Walden; or, Life in the Woods* [1864]. New York: New American Library, 1960.

——— *A Week on the Concord and Merrimack Rivers* [1839]. New York: Literary Classics of the United States, 1985.

Torrey, B., and F. H. Allen (eds.). *The Journal of Henry D. Thoreau.* New York: Dover Publications, 1962. (Reissue of 1906 edition published by Houghton Mifflin.)

Toumey, J. W. "What Ails New England Forests?" *Journal of Forestry* 26 (1928): 464–471.

Toumey, J. W., and R. F. Korstian. *Foundations of Silviculture.* New York: John Wiley and Sons, 1937.

Trefethen, J. B. "The Massachusetts Land and Its Wildlife: A History of the Residential and Migratory Game Birds and Mammals of Massachusetts." Master's thesis, University of Massachusetts, Amherst, 1953.

Vickery, P., and P. Dunwiddie. *Native Grassland of the Northeastern United States.* Lincoln, Mass.: Center for Biological Conservation, Massachusetts Audubon Society, 1998.

Webb, S. L. "Potential Role of Passenger Pigeons and Other Vertebrates in the Rapid Holocene Migrations of Nut Trees." *Quaternary Research* 26 (1986): 367–375.

Wessels, T. *Reading the Forested Landscape: A Natural History of New England.* Woodstock, Vt.: Countryman Press, 1997.

Whitney, G. G. "Fifty Years of Change in the Arboreal Vegetation of Heart's Content, an Old-Growth Hemlock–White Pine–Northern Hardwood Stand." *Ecology* 65 (1984): 403–408.

——— *From Coastal Wilderness to Fruited Plain: A History of Environmental Change in Temperate North America from 1500 to the Present.* Cambridge: Cambridge University Press, 1994.

Whitney, G. G., and W. C. Davis. "From Primitive Woods to Cultivated Woodlots: Thoreau and the Forest History of Concord, Massachusetts." *Journal of Forest History* 30 (1986): 70–81.

Williams, M. "Clearing the United States Forests: The Pivotal Years, 1810–1860." *Journal of Historical Geography* 8 (1982): 12–28.

——— *Americans and Their Forests: A Historical Geography.* Cambridge: Cambridge University Press, 1989.

Wilson, A. *American Ornithology; or the Natural History of the Birds of the United States,* vol. 5. Philadelphia: Bradford and Inskeep, 1812.

INDEX